6.50

INTRODUCTION TO ZOOGEOGRAPHY

JOACHIM ILLIES

Max Planck Institute for Limnology, West Germany

Translated by

W. D. WILLIAMS

Monash University, Australia

Macmillan

First published in West Germany by George Westermann Verlag 1971
Second German edition 1972

First published in the United Kingdom 1974 by
THE MACMILLAN PRESS
London and Basingstoke
Associated companies in New York Dublin
Melbourne Johannesburg and Madras

SBN 333 14383 3

Printed in Great Britain by
HAZELL WATSON AND VINEY LTD
Aylesbury, Bucks

Contents

Translator's Note v

Introduction vii

PART I ANIMALS AND THEIR ENVIRONMENT (ECOLOGICAL ANIMAL GEOGRAPHY) 1

Abiotic Factors 1

Individual Factors: temperature; moisture; light; current; oxygen; further factors 2
Ecological Valency and its Designation 14
The Network of Factors (Biotope) 16

Biotic Factors 17

Individual Factors: food; enemies; competition 17
The Network of Factors (Biocoenosis) 22

The Ecosystem 23

The Bioregions and their Fauna 26

Tundra 26
Taiga 29
Deciduous Forests 30
Tropical Rainforests 31
Sclerophyll Forests 32
Steppes 33
Deserts 34

Littoral Areas 35
Sea 36
Inland Waters 37

PART II ANIMALS AND THEIR DISTRIBUTION (HISTORICAL ANIMAL GEOGRAPHY) 43

The Faunistic and Geological Basis 43

Basic Principles of Causal Zoogeography 48

Faunal Regions of the Continents 53
Holarctica 55
Ethiopian–Oriental Regions: Ethiopian Region; Madagascar (Malagasy); Oriental Region; Wallacea 63
Neotropis 71
Notogaea: Australia; New Guinea (Papua); New Zealand; Oceania 75
Antarctica 85

The Faunal Regions of the Sea 87

Animal Migrations 90

Animals and Man 93
Hunting 94
Domestic animals 97
Man as a supra-organic factor 100

Bibliography 107

Index 110

Translator's Note

There is no modern comprehensive account of zoogeography written in English that approaches the subject in the way, and at about the same intellectual level, as does Professor Illies in this book. It was primarily for this reason that this translation was undertaken. Its availability will, it is hoped, fill the needs of those zoology students requiring an up-to-date 'overview' of the subject but who have neither the time nor, perhaps, the inclination to obtain this from the longer and much more specialised texts now available, or from reference to original primary sources. As a biologist who has himself contributed significantly to the field of zoogeography, Professor Illies is well able to provide, in an authoritative way, such an 'overview'.

A not unimportant secondary reason for this translation is that throughout the text there are various indications of the differences (or emphases) of approach which – though minor – distinguish it as the work of a European zoologist. In a world where English is increasingly becoming the dominant scientific language, English-speaking zoology students tend, I believe, to undervalue the different approaches adopted or emphases given in countries where English is not the native language.* If this translation provides a gentle reminder to such students that English-speaking zoologists have no exclusive rights to the way zoogeography is approached, and underscores the contributions

* It may be noted that the same comment sometimes applies even at higher academic levels; the delayed recognition by English-speaking zoogeographers of the importance of Willi Hennig's ideas on 'phylogenetic systematics' is a case in point.

from European zoogeographers, its production will have been doubly justified.

As all who command more than one language know, direct word-for-word translation from one language to another is not the most successful way of transferring ideas between languages. The present translation, therefore, while adhering as closely as possible to the original German text, does not fail to diverge from it when it has been thought that a certain idea or concept could be expressed more accurately in English in a quite different way. The translation is of the second revised German edition published in 1972 (*Tiergeographie*). The original edition (*Einführung in die Tiergeographie*) was first published in 1971. Various minor errors that occurred in the second German edition have been corrected in this translation.

Grateful acknowledgement is made of the help of Drs Peter and Heide Zwick (Limnologisches Institut, Schlitz, West Germany), who meticulously checked the accuracy of the draft translation.

W. D. Williams

Melbourne
July 1973

Introduction

The science of the distribution of animals is called zoogeography, implying that the subject matter of one discipline (zoology) is considered from the viewpoint of another (geography). However, zoogeography not only links these disciplines, but since it is a truly interdisciplinary science it has a further connection with a third discipline: ecology, the study of organisms and their environmental relationships.

The basic question asked by zoogeographers is: where does a specific animal occur, and why just there? We have to know therefore something about its area of distribution, although this raises a methodological difficulty in deciding what comprises the 'area of distribution' for any one species. Is the answer, for example, South America, in a geographical sense; or, from the habitat viewpoint, subtropical savannahs, or, considering its ecological occurrence, in rivers? A purely zoological answer can also be given: for a parasite, for example, it may be in the plumage of humming birds. These distinctly different statements about the distribution of an animal indicate the wide basis from which questions arise and the numerous ways in which they may be considered, all implicit in the single and comprehensive term, zoogeography.

With regard to the variety of possible ways the subject can be discussed, Hesse's basic work in 1924 drew a now commonly accepted distinction between ecological and historical animal geography. Both directions are followed in the present text; only thus can a comprehensive idea of this science be obtained. A final chapter is devoted to the relationships between men and animals, in so far as these are of zoogeographical significance. Lack of space demands a strongly condensed presentation, but if further details are required, the appendix gives a list of references to the literature.

PART I

Animals and Their Environment

(Ecological Animal Geography)

According to a definition given by HAECKEL (1866), ecology is 'the total science of the relationships of the organism to the surrounding environment, within which we can include in a further sense all conditions for life'. This definition has undergone many modifications. On the one hand these have resulted in its limitation to biological subjects (for example, the 'existence-ecology' of KROGERUS), and on the other in its wide application to a general ecology of 'the economy of nature' (THIENEMANN and FRIEDERICHS) wherein all scientific disciplines have a place. Generally, a position between these extremes is adopted, and ecological investigations are taken as concerning the environmental relationships of organisms. By 'environment' we do not mean the total surroundings of the organism, but follow UEXKÜLL in considering it as only those factors demonstrably possessing a direct relationship to the organism, such as climate, soil, feeding relationships, competitors, predators and so on. If relationships which influence single organisms are being investigated, we talk of autecology; if interactions within a population, of demecology; and if relationships between the environment and whole biotic communities are being considered, of synecology.

ABIOTIC FACTORS

The totality of physical and chemical factors presented to the organism as the local climate can be thought of as a complex or network of factors permitting the existence of the animals involved in that particular place, but, at the same time, limiting their area of distribution to it. Apart from those cases of entirely insular distribution and isolated montane distributions, the spread of an animal species over its inhabited region is opposed less by geographical phenomena than by climatic ones. In addition, at the edge of the zone of greatest adaptation, there is competitive pressure from other better adapted forms. In this context, climate must be regarded not only as the meteorological climate of the area in question, but as a microclimate only indirectly dependent upon the gross climate. It is this microclimate which occurs in the actual habitat of the species, for example in the soil, in a lake, in the foliage of a tree, or within an ant nest. All microclimates are composed principally of a series of elements which can be considered individually as separate factors in their influence on the spatial distribution of animal species.

Individual Factors

Temperature

Temperature, together with humidity for terrestrial animals, forms a so-called dominant factor, that is to say one of decisive importance for the existence of animal species. For a great number of species, persistence within a possible habitat is determined on the one hand by the average monthly temperature of the habitat (particularly for aquatic and soil inhabitants) and on the other by the temperature extremes, especially those of the winter months. Since the speed of physiological processes in organisms takes place according to the general law of chemical reaction rates (Arrhenius equation), an elevation of the environmental temperature of about 10°C results in a doubling of the metabolic rate. One can see, therefore, why the lives of cold-blooded animals in particular, that is those animals lacking the characteristic thermoregulatory mechanisms of birds and mammals, are directly affected by the temperature of their environment. Thus in many cases, annual isotherms, especially those for frost, determine the distributional limits of individual animal species or whole groups.

For active life, the temperature tolerance limits of organisms are −1.5°C and 80°C, the latter value applying to some bacteria and blue–green algae. For active life in animals, the highest temperature tolerated is about 55°C.

Upper limits at which animal life has been observed in thermal springs

Insects (several water-beetles), watermites and rotifers	<45°C
Microcrustaceans (e.g. *Cypris balnearia*)	<51°C
Chironomid midges (e.g. *Dasyhelea terna*)	<52°C
Aquatic snails (e.g. *Bythinia thermalis*)	<53°C
Protozoa (e.g. *Hyalodiscus*)	<55°C

Some species possess the capacity to survive well beyond these temperatures in a state of anabiosis, as resting stages, cysts, eggs and so on. The almost complete removal of water from the tissues in such a condition permits without harm a depression of the temperature down to −196°C and an elevation of it to over 100°C. This can be done experimentally with the African chironomid midge *Polypedilum vanderplanki*. The extreme capacity for temperature tolerance following desiccation allows the survival of the larval form of the midge in dried-out rock-hollows and rock-pools; the active life of the organism stands still, as it were, to be continued only after water renewal following the next rain.

The lower end of the temperature scale, particularly the freezing point, places special restrictions on animals. Most species, therefore, are excluded in their distributions from areas with low winter temperatures; species numbers decrease markedly from the tropics towards the polar regions. Geologically ancient groups are particularly sensitive, that is to say are particularly dependent upon heat: coral reefs disappear as soon as the temperature of the sea falls below 20°C; crocodiles have a lower temperature limit of +4°C. Because of the anomalous thermal characteristics of water, temperatures never fall below +4°C in the deeper layers of standing waters in temperate latitudes. Lake bottoms, therefore, remain permanently free of ice, and many cosmopolitan forms occur in the plankton as a consequence. In the terrestrial fauna, on the other hand, there are very few such cosmopolitan forms (apart, of course, from parasites and synanthropes).

The adaptation to a specific temperature optimum and the extent to which this optimum can be transgressed on either side of the temperature scale is rather variable. Several water snails (for example, the liverfluke snail *Limnaea*) occur in cold and periodically frozen pools as well as in thermal waters. Other forms, such as warm-water ornamental fish, tolerate only small fluctuations in temperature. According to the nature of the temperature tolerance and preferred temperature, we can distinguish between *eurytherms*, or widely tolerant forms, and *stenotherms*, species adapted to only slight temperature fluctuations. Examples of cold-stenotherms are provided by many inhabitants of cold springs and arctic soils; examples of warm-stenotherms are many species of tropical fish and birds. The classical example of a limited, temperature-dependent distribution within a narrowly defined area is given by flatworms in the upland streams of central Europe (figure 1). Here, in

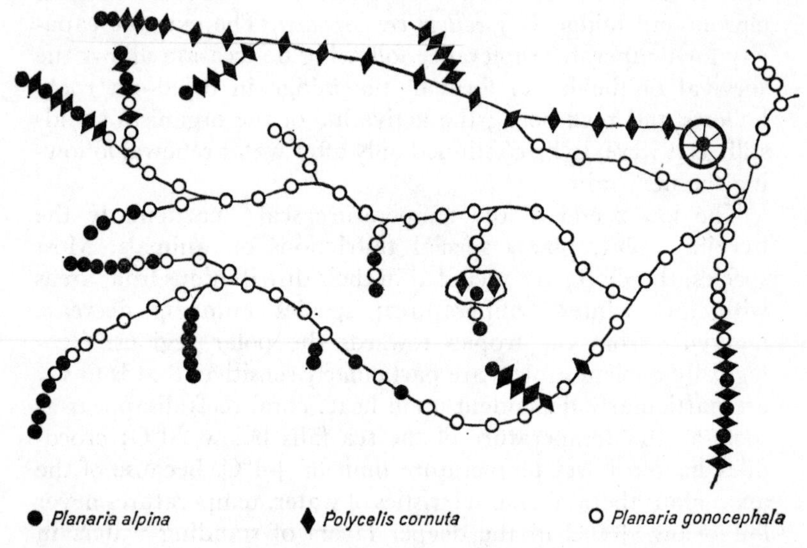

Figure 1 The classical zonation of flatworms in upland streams. Notice the influence of the watermill on the immigration of *Planaria gonocephala*

regular sequence downstream from the source springs, one can find the cold-stenothermal *Planaria alpina*, then the less stenothermal *Polycelis cornuta*, and then, in the lower courses, the eurythermal *Planaria gonocephala*. Still further downstream, these flatworms are followed by the extremely eurythermal *Planaria lugubris*.

In the case of warm-blooded animals (birds and mammals), the appropriate high temperature for vital functions is produced internally by the animal itself. In this way these species are to a certain extent independent of environmental temperatures, and can thus colonise even polar regions (polar bears, penguins, seals, etc.). However, the provision of warm blood requires much energy which must be obtained by additional feeding or by economies of growth (prolonged sexual maturity). Further, in times when little food is available, hibernation becomes necessary. Thus, in warm-blooded animals, too, there is a distinct, although indirect, independence upon the environmental temperature of the habitat.

Several relationships have regularly been observed:

1. *Size Rule* (*Bergmann's Rule*): In colder climates, representatives of warm-blooded animals are larger than representatives of the same species in warmer regions.
Examples are provided by the brown bear, red deer, wild pig (in Siberia this is more than double its size in Spain), fox (introduced examples in Australia attain there only half the body size of their English ancestors).
2. *Proportion Rule* (*Allan's Rule*): Extremities and other body appendages (tail, ears) are longer and larger in warm-blooded animals in warmer regions than they are in representatives of the same species in cold climates (a defence against heat loss).
Examples: foxes, lynxes, wild cats.
3. *Pigmentation Rule* (*Gloger's Rule*): Races of warm-blooded animals are more darkly coloured in moist and warm regions than races in colder and drier regions (as a result of more marked pigment formation to give protection from the light).
Examples: wolf, fox, hare.
4. *Heart-weight Rule* (*Hesse's Rule*). The need for the production of a greater temperature difference from the environment

causes the volume and weight of the heart of animals in colder regions to become distinctly greater than in races in warmer regions.

Moisture

Because of the great significance of water in the maintenance of active life, environmental moisture is a decisive factor in determining animal distributions. For most terrestrial animals, not only the availability of drinking water but also a certain degree of humidity in order to minimise water loss via transpiration is a vital necessity.

For aquatic animals, permanent water in their habitat is the most basic provision for existence; even short-term departure from their medium is impossible. This holds true especially for the primary aquatic forms, in other words those that are phylogenetically old, such as many Protozoa, Coelenterata, Echinodermata and fish. However, when the body possesses an efficient protective device to prevent desiccation (as in, for example, shell-bearing molluscs), short-term survival during desiccation and colonisation of marine tidal zones is possible. Finally, many aquatic forms possess special anatomical and physiological adaptations which protect their body respiratory surfaces (for example, gills) against drying out and enable them to leave the water. Amongst fish, the eel has the ability to crawl over short tracts of land in order to seek out new aquatic habitats; and amongst crustaceans, the Indo-Pacific 'coconut crab' *Birgus latro* is completely terrestrial and lives in the foliage of marginal trees.

With regard to terrestrial forms, some need a high degree of vapour saturation in their habitat because their skins do not have any protection against evaporation. The animals of moist, warm, tropical jungles, of littoral regions and of caves belong to this group. So do most soil inhabitants, for example the earthworms, and nearly all amphibian groups (frogs, newts, toads). Slugs and most thin-skinned insect larvae are likewise dependent upon permanently high humidity.

Animals living in dry places, in contrast, have the ability to eliminate evaporative water-loss so effectively that a temporary or permanent sojourn in a dry atmosphere is possible. Decrease in the outer surface area of the body (in-rolling) and a decrease in the rate of metabolism (anabiosis, encysting, etc.) are adaptations to aridity which are in particular displayed by the inhabitants of savannahs and deserts. Inhabitants of dry wood (death-watch beetles) and desert sand (many Tenebrionidae) display the most extreme adaptation: they have the ability to satisfy their water needs entirely by metabolic water produced

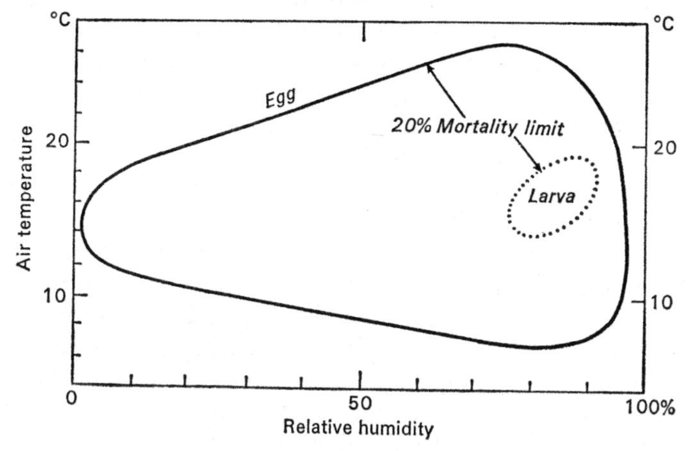

Figure 2 Combinations of humidity and temperature that can be tolerated by the larval and egg stages of the moth, *Panotis flammea* (after EIDMANN, 1941). The 20 per cent isomortale is shown, i.e. the boundary beyond which more than 20 per cent of animals die

by chemical transformation from food. In this way, several pests of stored food products (grain beetles, mealworms, moths) are independent of environmental water.

Tolerance to moisture or environmental aridity is closely related to temperature; mostly the need for moisture increases with increasing temperature. Moreover, the various developmental stages of a species often have very different moisture and heat tolerances (figure 2). The following ecological rule applies:

The development of a species in any habitat is limited by that factor which approaches the minimum for the developmental stage with the lowest ecological valency* for it. (*Liebig–Thienemann's law of the minimum*).

Light

During the process of assimilation, green plants use light energy in the formation of carbohydrates; that is, they store energy. All animal life is dependent upon the subsequent release of this energy; thus, because of its ultimate dependence upon plant products, animal life is impossible in the absence of light. Animals cannot use solar energy directly, although it is worth mentioning that several species of ciliates and hydrozoans accommodate as 'guests' in their body-cells unicellular green algae (*Chlorella*) upon which they may live. Apart from this sort of situation, direct sunlight is, indeed, injurious to animals cells, mainly on account of its ultraviolet component; pigment layers are therefore necessary to protect the outer skin of the animal. Conversely, those animals which live in habitats where light is absent (in groundwaters, the soil, inside plants, or internally in other animals) are mostly unpigmented. They are thus colourless or white and cannot withstand exposure to daylight. Earthworms that have crawled out of the soil on to the surface following a shower of rain die after only a short exposure to the direct rays of the sun, a phenomenon known as 'light-death'.

The intensity of the light preferred by animals for active life is very variable. Apart from nocturnal species (owls, bats, hedgehogs, etc.), there are some which are active at dawn or dusk (for example, many songbirds sing in the morning, and many insects swarm in the evening), and yet others which only attain their full activity in bright daylight (for example, lizards, many butterflies, jewel beetles, etc.). Light also plays an indirect but important role as a prerequisite for the functioning of the eyes; daylength has therefore some significance in determining animal distribution. Thus, some warm-blooded animals (birds, mammals) can only obtain the large quantities of food necessary for their thermal budgets if a sufficiently long day is available. As a result, many tropical songbirds cannot be kept in captivity in winter in the higher latitudes; they starve if not given, by artificial illumination of their cage, the necessary period

**Translator's note:* See p. 14 for a definition of ecological valency.

of light for feeding. The smaller quantity of winter light in the higher latitudes can, in this way, be a factor limiting distribution.

Animals living in darkness can sometimes produce their own light; this light mostly serves as an optical signal for species to facilitate meetings of the sexes (glowworms). Amongst deep-sea inhabitants, there are some fish species equipped with light-producing organs. Some of these fish use light to lure their prey, and almost always these organs can be switched off so that a rhythmic light pattern can be produced. A cave-living midge (*Arachnocampa luminosa*) in New Zealand has a larva that spins a web on the cave ceiling and attracts insects into its net by means of a light-producing organ in its body.

Current

The influence of the wind on the distribution of many animals is considerable, and especially for insects, spiders and protozoans. Active, flying insects as well as spiders passively propelled in webs may reach new habitats by using wind currents. Wind transport is particularly important in the initial colonisation of freshly created biotopes (islands). Such transport is effective over great distances: Gressitt (1961) was able to demonstrate, with the help of nets installed on ocean-going ships and aircraft, the presence in the southern Pacific of an aerial plankton in wind currents more than 1000 km from the nearest land (it included collembolans, termites, cicadas, mosquitoes, moths and small spiders). Cyclones, whirlwinds and hurricanes can also carry larger organisms to great altitudes and for many kilometres over the earth. Thienemann reported several cases from central Europe where even frogs and fish could be shown to have been spread and to have colonised isolated and otherwise completely inaccessible waters in this way. Historical documents from the Middle Ages had already reported on the descent of such 'rains of fish'.

On the other hand, for the inhabitants of isolated habitats the possibility of wind transport presents a danger: an aquatic insect that has been removed from its habitat by the wind has little chance of encountering a new, suitable one. As a consequence, selection by wind drift in extreme localities (especially high mountains above the tree line) has often led to the development of short-winged insect populations. On the Kerguelen Islands, for example, there are flies, butterflies and other insects which are completely wingless and which move entirely by running.

Directional marine currents are of great significance for the distribution of marine animals, especially planktonic larval stages. In the Gulf Stream, for instance, eel larvae are propelled to the European coast. Vertical currents which bring enriched deeper waters to the upper surface are important for plankton production and are necessary for the occurrence of rich marine fish populations. Oceanic plankton species exhibit considerable local differences as a result of such marine currents and so one finds that there are not only rich fishing grounds (for example, the north-west Atlantic) but also pronounced 'deserts' with no fish production worth mentioning and recognisable by the blue (plankton-free) colour of the water.

In fresh water, current is particularly important in running waters as a factor controlling distribution. Special adaptations (suction cups, hooks, bristles, adhesive glands, etc.) allow many 'rheophilous' species to defy the strongest current and colonise such habitats as mountain streams and running springs; several aquatic insect larvae, as well as some tropical fish and frogs, are so adapted. Convectional currents (up-welling of water layers as a result of different temperatures and densities) play a decisive role in inland lakes for they enable the plankton to remain in the illuminated zone of the epilimnion (see p. 41). Water currents are also significant as a means of transporting food and oxygen, especially in rivers. Motile food particles are intercepted using special methods (nets, filters, sieves and similar apparata) by such filtering organisms as sponges, and *Hydropsyche* and *Simulium* larvae. Other organisms produce the necessary currents for food and respiration for themselves. They do so by creating directional water currents through undulating body movements (many worms and insects) or through special current-producing organs (ciliary funnels, gills, gill covers).

Oxygen

The concentration of oxygen normally presents no problem for terrestrial animals, since any oxygen consumed is quickly renewed by the atmosphere and by the photosynthetic activity of the surrounding vegetation, which also removes the carbon dioxide produced by respiration. Only in caves, especially fumaroles (e.g. the Grotta del Cane at Pozzuoli), does a shortage of fresh air occur or is oxygen so lacking that animal life becomes impossible.

Except in the deeper layers, the quantity of dissolved oxygen needed by the comparatively sparse animal population of the sea is also always adequate and is renewed by diffusion from the upper surface. In fresh waters, on the other hand, conditions for the development of an oxygen insufficiency may easily occur, and these can occasionally lead to catastrophes (fish deaths). The temperature-dependent solubility of oxygen is inversely correlated with the physiological needs of aquatic forms; thus, warm water at a time of increased need has a diminished concentration of oxygen, and cold water at a time of decreased need, an increased concentration. (At 0°C, 1 litre of water contains 9.7 cm^3 0_2, but at 30°C, only 5.4 cm^3 0_2.) Extreme situations mostly occur therefore in mid-summer and at times of low water; they are exacerbated by the introduction of waste water as this causes additional decay processes which consume further oxygen. In the depths of lakes during the summer months, an oxygen deficit or a complete loss of oxygen occurs, leading to the formation and deposition of black muds containing hydrogen sulphide. It is only at the full spring or autumnal circulation that such lakes once again contain oxygen in their deeper layers. At the bottom of deep tropical lakes (for example, Lake Toba in Sumatra) there is no oxygen and complete circulation never occurs; this sort of biotope is therefore azoic, that is to say entirely lacks animal life.

Further Factors

Apart from the above-mentioned abiotic factors of general significance, other factors can be responsible for controlling distribution in some habitats. The most important of these should be mentioned here.

Water salinity is a factor determining the osmotic balance of the body, and this explains why – particularly among marine forms – only slight variations are tolerated. Thus, reef-building corals are so sensitive to any dilution of their milieu that ring atolls and barrier reefs show gaps where rivers flow into the sea. Only a few polyhaline organisms (for example, the brine shrimp *Artemia*) can withstand any upward fluctuation from the normal salinity of the sea (33–7‰). Downward fluctuations are equally inimical to life; brackish waters, being areas where marine and

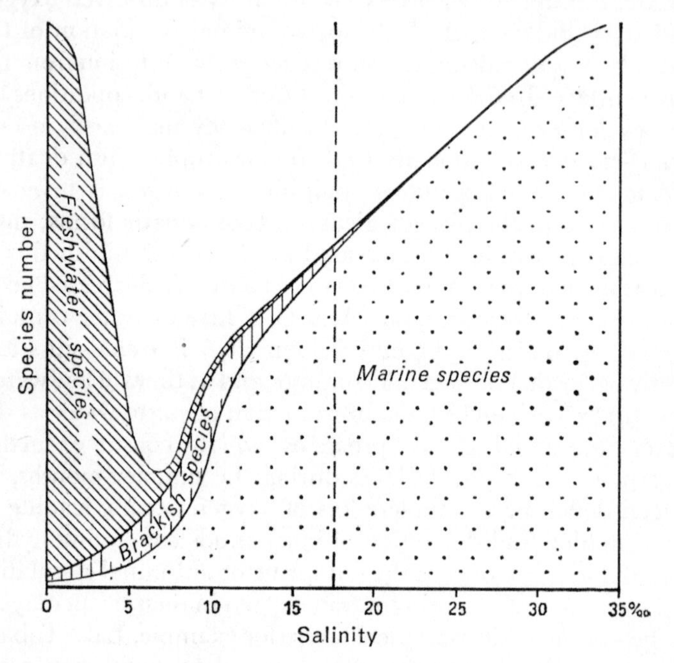

Figure 3 Species numbers of marine animals in relation to salinity (after REMANE, 1934). The minimal number is not in the middle, but is clearly displaced towards the region of lower salinity.

fresh water mix, are colonised by only a few euryhaline species (that is, salt-tolerant forms and forms tolerant of dilution) and exhibit an extreme, species-poor and characteristic fauna (figure 3). The Baltic Sea provides the classical example of the way in which salt concentration in the sea effectively controls distribution; only about 12 of the approximately 200 crustaceans of the Kattegat penetrate as far as the Gulf of Bothnia where the water is less saline. The colonisation of fresh water is an evolutionary process which makes high demands on the physiology of water removal from the cell (by pulsating vacuoles, kidneys) and it became possible therefore only late in the course of animal evolution. Consequently, limnic forms (freshwater inhabitants) are always descendants of marine or terrestrial ancestors.

Calcium concentration can also be of distinct significance for freshwater populations. Calcium-poor waters ('soft' waters), such as those streams on very old rock strata, develop a rich moss vegetation and as a result are particularly suited for animal colonisation. Nevertheless, such waters are avoided by, for example, the freshwater mussel (*Margaritana margaritifera*). Calcium-rich waters ('hard' waters), conversely, are clearly preferred by several stream insects (for example, the beetle genus *Riolus*). Acidity, increased concentrations of humic material, and other departures from the normal chemical composition of streams diminishes the number of inhabitant species; streams running through coniferous forests, for example, are almost unpopulated because they contain a dissolved extract derived from conifer needles. In lakes, water chemistry influences stratification and is thus significant for planktonic animals.

The influence of the moon must also be mentioned. In marine tidal zones, the periodic exposure to air when the tide is out promotes special adaptations in benthic organisms. However, some animals also display periodic phenomena in the depths of the sea and even in fresh waters, and these phenomena must be caused by the moon's influence. The activities involved are mostly migratory or swarming movements at certain phases of the moon. For example, worms of the genus *Eunice* (the 'palolo-worm'), polychaetes which live in the Pacific and bore into coral colonies, cast off the hinder part of their body at a certain phase of the moon; the cast-off portions are filled with sexual products. In Heligoland, the midge *Clunio marinus* regularly swarms in the tidal shallows at the syzygy of the moon. In Lake Victoria, Africa, the tropical mayfly *Povilla adusta* regularly hatches every four weeks on the second day after the full moon. And the activity of eels moving downstream in rivers, as indicated by the numbers caught by an eel-shocker, regularly increases during the last quarter of the moon to many times its normal extent (figure 4). All these phenomena cannot yet be

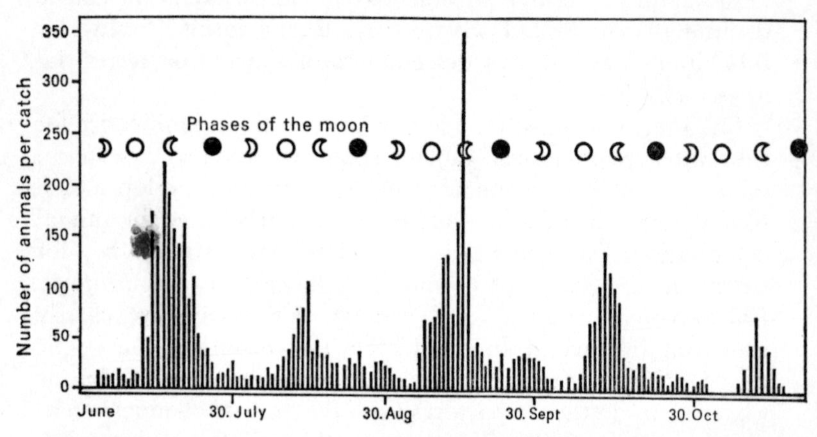

Figure 4 The migratory activity of eels and phases of the moon; daily numbers of eels caught in the upper Rhine and the phases of the moon (after JENS, 1953).

explained successfully, but the 28-day lunar rhythm is obviously fixed in many body functions, and in humans is demonstrable as a hormone cycle.

Ecological Valency and its Designation

The double concept of eurytherm–stenotherm for the different grades of tolerance to temperature has been used in the treatment of a single factor. But many more systems of naming such valencies can be found in the literature (for example, -xen, -phil, -biont; oligo-, poly-; rheo-, thermo-, oxy-, etc.) which greatly increase the layman's difficulties in understanding ecological texts. Figure 5 defines these designations and shows how they should be consistently applied (after VOUK, 1939). For each factor an oligo-, meso- and poly- range is differentiated; in the case of oxygen concentrations, for example, small, normal and excessive amounts of dissolved oxygen could be considered. All organisms which are restricted to one of these ranges are stenoecious and may be characterised using combinations of the prefix sten(o) and oligo-, meso- or poly- in front of the Greek word for the factor being considered (therm, oxybiont, rheob,

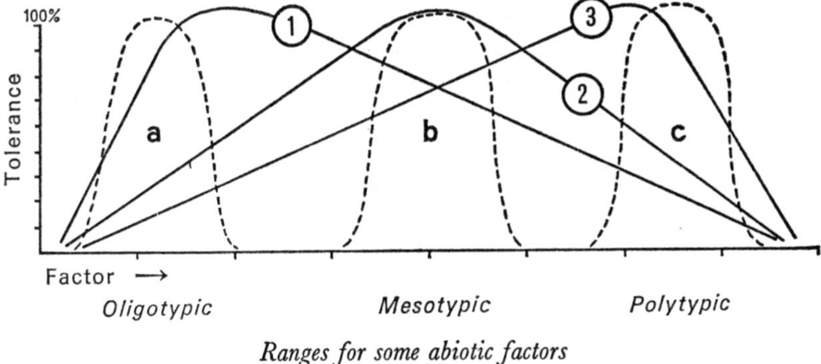

Ranges for some abiotic factors

	Oligo-typus	*Meso-typus*	*Poly-typus*
O_2 concentration (mg/l)	0–4	4–7	>7
Temperature (°C)	10	10–20	>20
Salinity (‰)	0–30	30–35	>35
Current (m/s)	0–0.05	0.05–0.5	>0.5

Examples of the scheme's application:

(a)	Oligo-steno-therm	Glacial flea	(*Isotoma saltans*)
(b)	Meso-steno-rheob	Barbel	(*Barbus fluviatilis*)
(c)	Poly-sten-oxybiont	Midge	(*Liponeura cinerascens*)
(1)	Oligo-eury-oxybiont	Chironomid	(*Chironomus plumosus*)
(2)	Meso-eury-rheob	Freshwater limpet	(*Ancylus fluviatilis*)
(3)	Poly-eury-halob	Jelly-fish	(*Aurelia aurita*)

Figure 5 Ecological valency: its designation, amplitude and the position of the optimum (after VOUK, 1939).

etc.). Euryoecious organisms, which extend over greater ranges of valency, are designated by the prefix eury- in front of the respective factor. In this way, for each factor there are six possible grades of valency, as indicated in figure 5.

The Network of Factors (Biotope)

Each abiotic factor makes specific demands on the animal's adaptations, and each of the six valencies indicated in figure 5 can be applied to every one of the factors. In spite of this, no factor operates independently, and there are always more or less distinct and direct relationships between them: for example, temperature and humidity, temperature and oxygen concentration in flowing media, and salinity and density in the sea. Such factors are physically correlated with each other and occur in quite specific combinations in nature.

Considering the network of those abiotic factors that constitute for any animal the climate of the environment in a given locality, quite characteristic combinations frequently seem to occur. Ecology at this level ('coenographic') is concerned, therefore, with the extent to which forms are either stenoecious or euryoecious. In other words, it is concerned with the narrowness or breadth of adaptation within the spectrum of factors that occur in a quite specific locality, the biotope. Thus, desert animals are adapted, for example, to high temperatures and low humidity or temporary aridity, cave inhabitants to darkness, dampness and cold and inhabitants of the deep sea to high pressure, low temperature and darkness.

On the basis of the physiological adaptations displayed by animals, it is possible to give, conversely, the following definition: in the locality of maximum occurrence, all abiotic factors are at the optimum of tolerance. This optimum is also called the habitat of the species. At the same time such habitats (for example, leaves of a tree, herb layer of a forest, humus soils, rotting logs, lake bottoms) form the structural component or substrate of a biotope, as for example in a deciduous forest, a field and so on. The habitats of single species do not always coincide with the distribution of the corresponding biotope, for overlapping associations can be found. Thus, a leaf-beetle which is confined to a certain tree species as a habitat will not only colonise this tree within the forest community but will also do so when the tree stands isolated in a meadow biotope or in a forest clearing. Similarly, a rodent (for example, a hare) feeding on the herb layer will not only eat grass in a field but will do so from the bordering forest too, and a soil-dwelling worm will occur in the soil of the forest as well as that of a field. It can be seen then that the extent to which animals are associated with individual biotopes varies, and it is necessary therefore to consider different grades of biotope association. Biotope-characteristic species (*indigenae*) are distinguished from visitors (*hospites*), neighbours (*vicini*) and passengers or stray 'guests' (*alieni*). It is also customary to name biotope associations with reference to the principles of plant sociology. According to frequency, the following are distinguished: 'exclusive' species which are wholly restricted to the biotope in question; 'preferent' species which prefer to occur there but do not do so exclusively; 'indifferent' species which frequently but not regularly occur; and, finally, 'strangers' which are species only occasionally to be found in the biotope. The first two groups are referred to as *characteristic* species as it is possible to characterise a given biotope with their aid (for example, 'oak–hornbeam forest', or, in zoology, the 'ruffe–flounder region' of a river). According to abundance, that is the frequency of individuals within a biotope, dominant species (each with more than two per cent of all individuals) and recessive species (less than two per cent) may be distinguished.

BIOTIC FACTORS

The living environment, the totality of animal and plant individuals each animal encounters in its area of occurrence, represents a network of factors that is of tremendous significance. Above all this applies to the vegetation of the locality, since plants are for all animals the source of oxygen, indirectly or directly the origin of food, and they often form the habitat too. Likewise, animals, as either predators or prey, parasites or competitors, have a direct influence on species populations. The sum of this influence constitutes biotic factors.

Individual Factors

Food

Because of the fundamental significance of food, numerous adaptations to specific methods of obtaining food can be identified. In a sense, every animal species can feed in one of several possible ways. As a result, the following general ecological rule applies: where two animal species occur together, there are at least two different feeding habits. It follows that a variety of systems becomes possible if one wishes to divide animals according to feeding habits, and each demonstrates in different ways the close relationship between feeding habits and distribution.

The type of food ingested depends on the sort of structural adaptations the animal has, the way it lives, and the nature of its habitat distribution. The first animals were planktonic feeders, and this method of feeding is still found above all in primitive marine animals where it is often aided by ciliary and filtering apparatus. However, even such advanced forms as the blue whale feed in this way. Periphyton feeders, also an ecologically old group, dwell amongst algal mats and are mostly smaller animals with scraping mouthparts (many snails and insect larvae). Animals that feed by sucking absorb fluid food and thus are parasites or plant juice feeders: mammalian juveniles belong in this category too. Mud on the sea-bed and terrestrial humus is eaten by animals which cause the substrate in which they move to pass through their bodies; digestible material is withdrawn from it in their guts. (The earthworm, *Lumbricus*, which attains densities of about four million per hectare in garden soils, provides an example of this mode of feeding.) Excrement, dung and carrion feeders fall into this category, for they also unselectively ingest the substrate, the food component being selected in the gut. The intake of soft food is common to all these types. A contrasting type is involved when hard food is utilised. Here, the problem is overcome by swallowing the food whole or in large pieces (snakes, predatory fish), or by chewing it down to suitably sized particles with appropriate mouthparts (teeth, grating plates, etc.). Almost all mammals belong to this last category.

According to the sort of food involved, one can distinguish between *herbivores* (plant eaters), *carnivores* (flesh eaters) and *omnivores* (generalised eaters), as well as further subdivisions in each of these groups (for example, carrion eaters, insect eaters, grain eaters, etc.). And according to food variability, one can distinguish between *euryphages* (for example, the wild pig whose food includes mosses, ferns, fungi, berries, worms, insects, mice, carrion and injured game) and *stenophages*, the latter being those animals closely tied to a specific food. *Monophages* represent an extreme case of specialisation; many parasites and insect larvae belong to this category. Most of our large butterflies, for example, are bound to one or a few food plants (the silk moth, for instance, feeds only on the mulberry tree).

Dependence of butterflies on food plants

Number of food plants	1	2	3	4	5	6	7	8	9	10
Number of butterfly species	132	53	57	38	23	15	14	11	5	9

Polyphages and *oligophages* can be distinguished by the quantity of food needed. Thus the polyphagous mole consumes 150 per cent of its body weight daily in feeding, whereas the mouse-eared bat consumes only about seven per cent. Feeding frequency also varies widely, and consequently so does the ability to survive periods without food. Freshly hatched ladybirds starve after only a few hours if they do not encounter any prey (aphids), and without food the larvae of the large water beetle (*Dytiscus*) starves in a day. Of the native mammals, the mole and shrew can only survive a few hours without food. On the other hand, most fish, for example, can endure periods of starvation lasting for many weeks. The Amazonian crocodile eats and is satiated only once a year, during the time of fish spawning. The bed-bug (*Cimex*) survives without food in a starved condition for a period as long as 10–18 years. Flatworms live off their own body tissues during periods of starvation, and specimens of *Planaria* can, in this way, revert to 1/300 of their original volume within a year.

Finally, it should be mentioned that many animals have different feeding habits at different times in their development. The larval stages of insects are often dependent upon sources of food quite different from those of the adult animals: for example, butterflies feed first on leaves, then on nectar; midges feed first on plankton, then on blood. Additionally, adaptation to specific feeding habits is not always restrictive, so that even in stenophagous animals there exists a potential for a much broader food basis. The contrary may also apply: the arctic fox can feed entirely on excrement, the donkey on paper (cellulose), horses on the Faroe Islands exclusively on fish and Mongolian horses on dried meat.

Enemies

It can be demonstrated that, as with feeding, each species also has a spectrum of relationships and adaptations which enable it to cope with a hostile environment. One expedient way of considering enemies is to divide them into predators and parasites, although this division is not absolutely clear-cut and there are many transitional forms.

Predators, that is to say carnivorous species feeding upon other animals which they hunt and destroy, exist in all biotopes, and they pose a constant threat for almost every organism. The smaller animal species (planktonic organisms, microscopic soil inhabitants and most insects) are especially exposed to and practically defenceless against attack from larger animals. However, even some of the larger animals have numerous predators; field mice, for example, figure in the diet of more than a hundred predators (including birds of prey, predatory mammals, wild pig and even such predatory fish as pike). Conversely, there are a few large animals – the whale and the elephant provide examples – which are quite without threat from predators in their natural habitat. Adaptations to predation include in particular an increased rate of reproduction, but there are additional protective devices (shells, scales, spines, poison glands, warning colouration, camouflage) which restrict the number of possible predators and contribute to the establishment of a predator–prey balance operating as a simple, self-regulating system. In the latter, increased rates of destruction of prey reflect on the predators, for they are then deprived of their food base and their own population falls; following this fall, the population of the prey can again increase. In this way, regular fluctuations in the population sizes of predator and prey may occur, as is shown particularly clearly by populations of lynx and hare (figure 6).

Parasites are animals which feed on other organisms but which

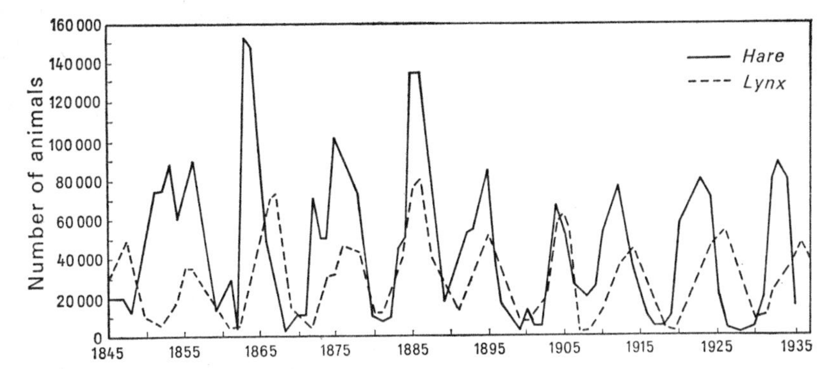

Figure 6 Changes in population numbers of the Canadian lynx and snowshoe hare (predator–prey), measured as the number of skins of both of these fur-bearing animals traded in Canada (after Odum, 1954).

do not kill them in so doing. Most parasites live in continuous contact with their 'hosts' which they harm either externally (*ectoparasites*, for example, fleas, lice, bugs) or internally (*endoparasites*) whilst feeding on the host's body tissues or gut contents (for example, intestinal worms). The animal involved protects itself against such enemies by special defensive mechanisms and physiological adaptations, although these in turn are overcome by parasites through specialised adaptations of their own. As a result, an attachment to a specific host often develops. The majority of parasites, indeed, are monophagous, that is host-specific to a certain species of animal or to a suitable group. Finally, it should be mentioned that a fluctuating balance similar to that shown in figure 6 for a predator–prey relationship develops between host and parasite. As before, an increase in the frequency of parasite attack harms the host population and thereby reflects deleteriously on the parasite itself.

Competition

Because of competition, the influence that related species have upon each other is of considerable significance. Wherever any limited basic resource becomes scarce because of simultaneous utilisation by the occupants of a habitat, the individuals are in competition, and the weaker suffer in proportion to the strength of the stronger. As the demands of conspecific individuals are particularly similar, it is these that experience the strongest mutual competition. Considering the immense reproductive capacity of many species and the fact that of the thousands of young animals (larvae) produced, usually only a few reach sexual maturity, and since enemies or disease cannot explain the deficit, it must be intra-specific competition which provides the most hostile of all biotic environmental factors in the struggle for existence by any individual animal species. Consequently, a single species can show regular fluctuations in population size due to competitive pressure, and a cybernetic system results wherein regular cycles of abundance and rarity succeed one another according to set principles (for example, 'cockchafer years' are regularly recurring mass population build-ups of this species every four years).

The mechanism of intra-specific competition, and the associated selection of inherited variations (mutations) which prove to be better adapted, forms the model developed by DARWIN (1858) to explain the origin and variation of species (evolution); it is thus of considerable biological significance. In any consideration of competition from an ecological viewpoint, we again find that there is an underlying network of abiotic factors, with food factors (the chief objective of competition) and predation factors (as negative selection) superimposed on it.

The Network of Factors (Biocoenosis)

Consideration of biotic factors indicates that they are functionally closely connected and can be understood better by discussing the combinations in which they influence animals in nature. For each animal species within its habitat, it is true that enemies (predators and parasites) exert pressure on a population, whereas food supports it. A labile balance occurs in the cybernetic systems of consumed and consumer, predator and prey, wherein the species occupies the most persistent position. Lateral pressure is exerted by competitors, that is to say neighbouring species (either geographical or ecological neighbours), and if these are superior they will replace the original species. The central position has been termed the ecological niche of the species. It is more than just the habitat (see p. 16): it is the optimum environment (considered as a combination of favourable abiotic and biotic factors) which both allows the existence of and is occupied by the species.

Species with ecological niches in the same biotope together constitute a *zoocoenosis* or, if plants are also considered (the *phytocoenosis*), a *biocoenosis* (biotic community). Smaller structural

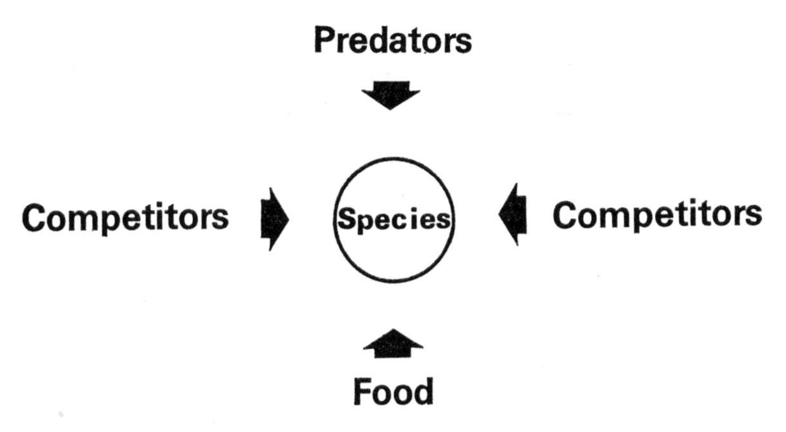

Figure 7 Position of the species within the network of biotic factors: the 'ecological niche'.

parts of a biocoenosis have been distinguished by some authors as *merocoenoses*. Biotope and biocoenosis thus correspond to a vessel and its living contents; together they constitute an ecological unit, the ecosystem.

THE ECOSYSTEM

Combinations of certain animals and plants in specific localities are natural phenomena; as ecosystems, biocoenoses in specific biotopes are essential structural components of that part of the earth's surface inhabited by living organisms. This ecological relationship was discovered by MOEBIUS (1877) in the oyster beds of Kiel Bay. He defined the biocoenosis as 'a community of living organisms comprising an aggregation of species and individuals that reflect the average external environment, that are mutually dependent, and that are permanently maintained in a given area'.

Modern ecology emphasises the fact that material and energy pathways occur in such ecosystems, and these provide the individuality and characterise the biotope in question. Autochthonous ecosystems have closed systems, that is to say are more or less independent of their surroundings (lakes, peat-bogs, jungles, coral reefs), whereas the material and energy systems of allochthonous ecosystems depend significantly upon inputs from other ecosystems (for example, rivers, the deep sea, caves, soil strata). Many ecosystems are unstable and, through succession, change in a regular manner during their existence (for example, ponds become filled in and finally converted to swamps, see figure 13). Other ecosystems are stable and represent the climax at the end of a succession (peat-bogs, deciduous forests). The following rules, the so-called biocoenotic principles, apply to all ecosystems:

1. *Principle* (THIENEMANN, 1918): The more variable the environmental conditions of a biotope, the greater the number of species of the associated biocoenosis.
(This applies, for example, in the insect merocoenoses of tropical jungles, where it is easier to find 100 different species than 100 examples of one species.)
2. *Principle* (THIENEMANN, 1918): The more that biotope environmental conditions diverge from the normal and optimum for organisms, the less the species diversity and the more characteristic is the biocoenosis (single species occur in even greater numbers).
Such species-poor biocoenoses include, for example, the tundra (figure 8); the faunas of polluted waters, salt lakes and of the ocean bottom are also poor in species and rich in individuals.
3. *Principle* (FRANZ, 1952). The greater the continuity of biotope environmental conditions during development, and the longer the biotope has remained homogeneous, the more diverse is its biocoenosis and the more well-balanced and stable it is.

(Coral reefs, which have high species diversities, provide examples of such old biotopes. The effect of this principle can be demonstrated in caves of different age.)

All three principles are concerned with the number of species in a biocoenosis. This number is greatly influenced by competition, but is also, on the other hand, an expression of the availability of a life-form type. The ecological applicability of the laws with respect to ecosystem saturation by different numbers of co-existing species is explained by the fact that species which are not closely related taxonomically (most are placed in different genera by systematists) can occur in the same biotope without displacement because of decreased mutual competition. Because of taxonomic difficulties (there is no universally valid definition of a genus), this empirical fact is not framed as a law as unequivocally as one would wish. Nevertheless, it occurs in many empirical rules which have the same ecological basis (Monard's principle, Gause's principle, Jordan's rule, the competitive exclusion principle):

Within a biotope closely related species do not usually occur in the same place, at the same time, or in the same habitat.

Thus, in a biotope there is often only one species from genera of high species diversity or from groups of species, and the generic coefficient (number of genera/number of species) is usually 1. In other words, many species in an ecosystem indicate many genera, and few species, few genera.

There is a clearer connection between biotopes and the many members of their biocoenoses with body forms resulting from morphological adaptation. Such ecologically conditioned similarities in quite unrelated animals have been designated *life-form types* (after Remane). The thin, worm-like bodies of the inhabitants of sand interstices, the well-defined form of the beak in many birds with specialised feeding habits, and the streamlining of mayfly larvae in running waters provide examples. These adaptations develop independently and by convergence in different families, so that biocoenoses are encountered in areas which are geographically quite distant and which as a result have quite different faunas, but which nevertheless have animals of similar body form. Biocoenoses with similar life-forms are termed *isocoenoses* (REMANE).

THE BIOREGIONS AND THEIR FAUNA

When considering the fauna and its associations (zoocoenoses), the vegetation was taken as part of the local conditions and the biotope. However, as plant associations (phytocoenoses) themselves directly depend on local conditions, the world's vegetational zones and their faunal associations may be regarded as providing the fundamental pattern for distribution. In the ecosystem concept, that is the concept of a unity of organisms and climatic factors confined to a particular locality, the connection between biotic and abiotic factors must be taken into consideration. It leads to the recognition of specific *bioregions* of the world, of areas having identical ecosystems with faunal and botanical isocoenoses in similar biotopes. (Although the concept of isocoenosis is referred to here, faunistic and geographical considerations are excluded for the present – they are dealt with in greater detail later, see p. 43.) The bioregions of continents, which in principle parallel geographical areas, reflect the fundamental significance of temperature; the vegetational zones encircling the world form this basic biogeographic pattern, and result from decreasing temperatures polewards together with regional variations in humidity. The geographical distribution of terrestrial bioregions is shown in figure 8. Figure 9 indicates the basic pattern of interrelationships between bioregions, climate and soil conditions.

Tundra

In the polar regions and on the higher parts of mountains with permanently frozen soil and a short summer growing season, there is within the 0°C annual isotherm a characteristic arctic or alpine vegetation (lichens, mosses, dwarf bushes) in which trees as a growth form are largely absent. Because melt-water on the frozen soil does not percolate away, pools, swamps and peat-bogs form. These extreme climatic conditions result (in

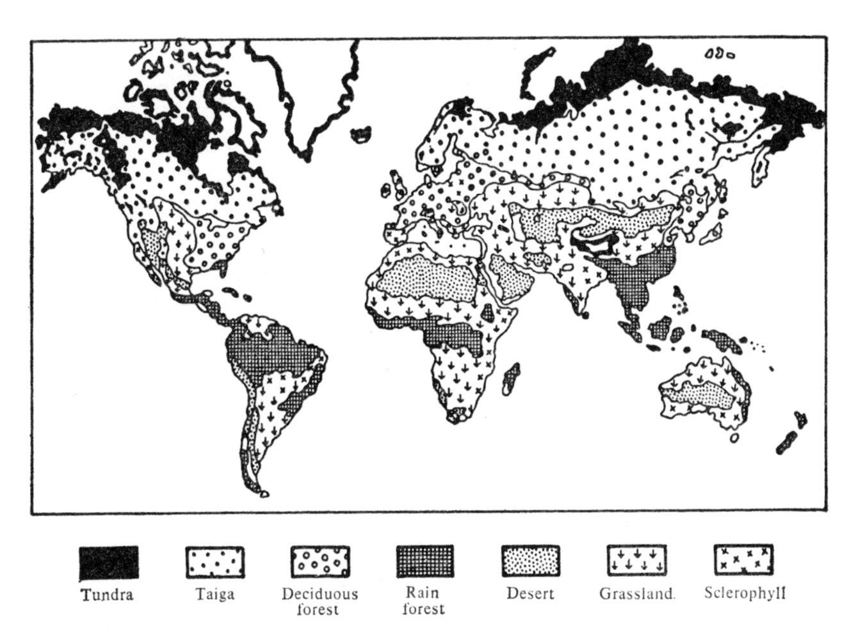

Figure 8 Bioregions of the continents (after TISCHLER, 1955).

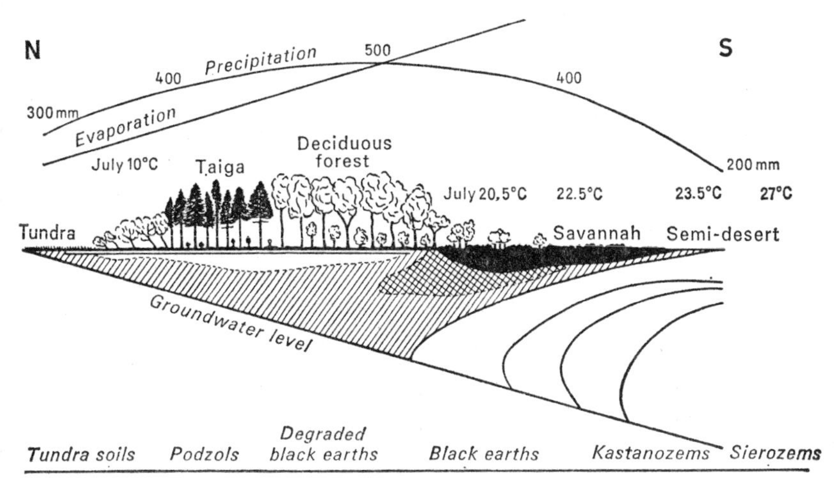

Figure 9 Interrelationships between bioregions, climate and soil conditions (after KASCHKAROW, 1944)

accordance with the second biocoenotic principle) in a biocoenosis of low species diversity. In the arctic tundra, and especially in the northern borders of Eurasia and North America, only a few mammals are represented: reindeer (*Rangifer*), polar fox (*Alopex lagopus*), alpine or polar hare (*Lepus timidus* and *L. arcticus*), musk oxen (*Ovibus* – now only found in Greenland and Canada), lemming (*Lemmus*) and several burrowing species of mice. The polar bear (*Thalarctos*) and arctic marine mammals can be found at the edge of the continental ice. Bird and insect faunas of the tundra are also poor in species. Physiological adaptations to the extreme climatic situation include special cold resistance, both day and night activity during the summer period, and bright body colours for camouflage.

Apart from the arctic (and antarctic) 'tundral' formation, a climatically comparable *oreal* (after De Lattin) occurs as an alpine formation above the treeline on high mountains. A special mammalian fauna is absent from these areas, which are mainly insular and small, although birds and insects similar

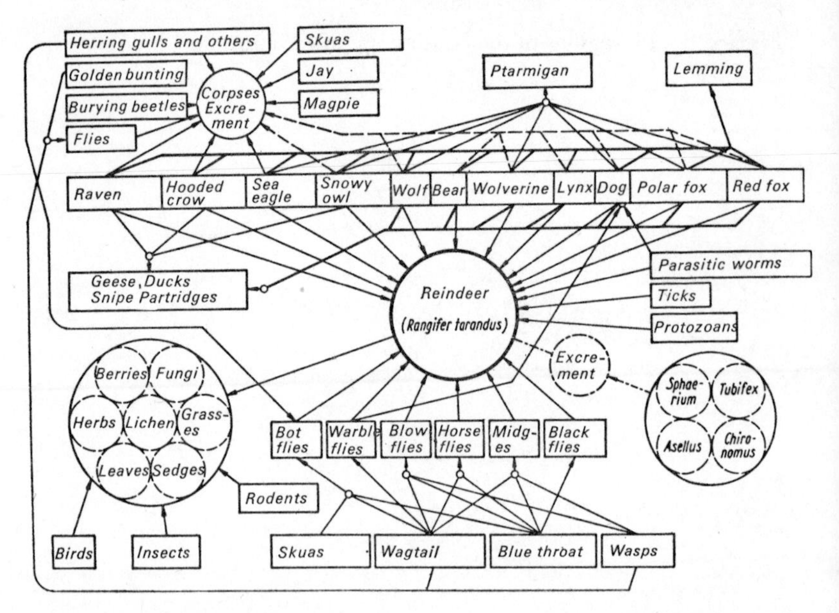

Figure 10 Trophic relationships in the tundra and taiga: those of the reindeer are emphasised (after Kühnelt, 1965)

to those of the tundra are represented, sometimes even as identical 'boreo-alpine' species (e.g. the ringed thrush, figure 17).

The tundra is a relict ecosystem and had a much greater distribution during the Glacial period. Part of its former fauna (mammoth, woolly rhinoceros) is extinct, and another part (musk ox, reindeer, polar bear) is clearly in the process of retreat. As a result of the extreme and hostile climate, the populations of such few members of the biocoenosis as remain are particularly sensitive to human interference. However, natural catastrophes (for example, the mass migration of the lemming) also basically express the unbalanced biocoenotic situation. The immense swarms of midges which characterise the tundra are a reflection of the presence there of a great number of suitable waters for the larvae (melt-water ponds) and the fact that females can live on plant juices and do not need to suck blood.

Taiga

The tundra connects southwards with the region of coniferous forest, which though not quite as extreme climatically is still characterised by a distinctly low temperature (figure 9). The fauna includes all the tundra animals, but now with the addition of most of the fauna of the forests of more temperate latitudes. The ecological relationships of reindeer indicate the most important animals of the region and their interactions as food, competitors, predators or parasites (figure 10). The predator–prey relationship of the lynx and hare (figure 6) also relates to this bioregion. Other large mammals such as the elk (*Alces*) and red deer (*Cervus*) occur here in a way similar to that of the reindeer.

On the floor of the coniferous forest a thin and not very productive layer of humus develops, harbouring a soil fauna which includes mites, spring-tails, nematodes and numerous insect larvae. In general, the insect fauna of the more diverse plant associations of the taiga is distinctly greater than that of the tundra. The extent of the taiga has not decreased following the Glacial period, and in area this bioregion is still the largest in the world. In large parts of Siberia and Canada it is also one of the regions least injured by man. Excessive logging and fur trapping, nevertheless, have already led to considerable harm. Hopefully, at least some areas representative of undisturbed taiga will be preserved in national parks.

Deciduous Forests

The region of temperate deciduous forests that are green in summer is typically well-developed in eastern North America and Central Europe. It is characterised by a diversity of ecological niches, and the number of faunal elements which occurs is large. To a great extent the fauna of the taiga is still present in the deciduous forest (for example, all of the large mammals and most bird species), but there are also animals with a low resistance to cold, insects associated with deciduous trees and the attendant flora, concurrent insectivorous birds, and so on. High summer temperatures allow animals from warmer southern regions to remain in the deciduous forest zone at least temporarily. And finally, the autumnal leaf fall leads to a deep layer of humus on the ground where an extremely rich soil fauna develops. Most animal groups display their greatest species diversity in this region (excluding the tropics).

Since deciduous forest areas are so climatically favourable, human populations are densest in this bioregion, and consequently it is in such areas that most damage to animal life has been caused by civilisation; and where the forests have not fully been cut down and replaced by pasture, there are plantations of commercially useful timber in monoculture. As a result, the fauna of deciduous forests now has many deficiencies and gaps; the aurochs, lynx, wild cat, wolf, bear, otter and beaver, for example, are now entirely or almost extinct in Europe. All other animal groups also show considerable loss as a result of civilisation. Sanctuaries can be significant in the conservation of small mammals, reptiles, insects and so on, but for the larger mammals and birds assistance can only be provided by large reservations and by enclosures and areas wherein re-establishment can be effected. It is only in the Carpathians and Urals that some undisturbed areas of European deciduous forest remain, but even there the outcome of the battle between natural protected areas and a combination of hunting and tourism is uncertain.

Tropical Rainforests

Moist tropical and subtropical forests with evergreen foliage and continuous growth provide maximal opportunities for diversification for many groups of animals. Thus, the greatest species diversity occurs in the tropical jungle for most animal groups, and, if bright colouration and bizarre appearance are considered, the most extreme evolution too. Birds and insects (butterflies above all) are particularly conspicuous because of their magnificent colours and great diversity of life-form types. The favourable climatic conditions, especially the high humidity and the permanent absence of frost, also allow the rich development of many archaic animal groups either absent from other regions or playing only a small role there: amphibians (the Apoda) and reptiles (pythons, turtles), slugs, land planarians, *Peripatus* (Onychophora) and jewel beetles (Buprestidae) are examples. In general, almost all insect orders in this bioregion contain additional and mostly very old families. With regard to parasitic hymenopterans (Chalcididae) and rove beetles (Staphylinidae) alone, thousands of tropical species are already known, and according to recent estimates these probably represent less than 50 per cent of the tropical species of these families that exist.

The number of individuals of a species, on the other hand, is usually small (in accordance with the first biocoenotic principle, see p. 24). Tropical rainforests, moreover, are distinguished by the speed of the circulatory processes of their material and energy pathways: hardly any humus is formed on the tropical jungle floor for almost all organic materials and essential minerals occur within the living system and on the death of this are immediately attacked by intense decompositional processes (natural waters in jungles are for this reason usually extremely deficient in electrolytes). As a result of this intensive metabolism – which continues without any reserves – animal production is small in tropical rainforests: the majority of mammals, birds and insects are 'rare', in other words they are encountered on only isolated occasions, and huge aggregations of animals such as are found in some aquatic ecosystems are almost unknown in the terrestrial ecosystem of the tropical rainforest. The low density of human populations in tropical jungles is partly explained by this poorness of animal production, especially if such populations are still at the lower human cultural stage of being collectors and hunters.

Inhabitants of the rainforest attempt to obtain space and inorganic nutrients for the cultivation of tropical crops (manioc, bananas, rice, etc.) by using fire to clear the forest. However, soil mineral deficiencies usually permit only a limited number of harvests, so that the settlers need to abandon the area after a few years and burn fresh clearings; in the subsequent ecological succession, a sparse secondary forest appears which is faunistically and floristically impoverished and which only gradually reverts back to climax associations. This bioregion, therefore, is very seriously threatened throughout the world. Moreover, political troubles in the tropical states make a fully responsible programme for conservation more difficult. Thus not only are the mammals and birds threatened by ruthless hunters, but also the other animals of the region are threatened by the destruction of the original forest.

Sclerophyll Forests

Lack of moisture, especially in the dry climates of the subtropical and temperate latitudes, favours the formation of more open, dry forests and sclerophyllous bush vegetation, for example the macchia vegetation (shrubs and low evergreen trees) of the Mediterranean area. In the middle latitudes, pine-heath forests and local bushy heaths are conspicuous. In accordance with the reduced variety of plant species, the fauna is also distinctly less diverse. There are transitions with the fauna of adjacent regions (deciduous forest and steppe) since areas of sclerophyll are often only small and occur as scattered islands, as it were, in climatically extreme parts of other bioregions. Because of the negligible value of sclerophyll forests as timber, they have been replaced by cultivated crops (olives, date palms). Furthermore, the artificial irrigation used in the cultivation of crops initially derived from deciduous forests (citrus fruits) can also exert a lasting change on the original character of the sclerophyll forest and thus on its fauna. In dry forests the low environmental humidity often causes forest fires (for example, in Australia), and these in turn cause great damage to the fauna. In sparsely settled areas, especially in the southern hemisphere, there still remain some places which represent undisturbed examples of natural sclerophyllous forest.*

Steppes

Savannahs and steppes are regions in which there is a complete cover of grass formed from 'hard grasses' interspersed in which are, at most, isolated trees, so that the landscape as a whole appears as an extensive grass plain. This sort of landscape develops wherever favourable temperatures for vegetation coincide with low rainfall and a deep-lying water table (figure 9), and it is particularly characteristic of subtropical and temperate dry areas on both sides of the equator. Rapidly moving animals predominate as the life-form type; hooved animals (cattle, horses, antelopes, giraffes, camels), rodents (rabbits, guinea-pigs, rats), the large cats (lions, cheetahs) and dog-like forms (wolves, hyenas, jackals). The locomotion of hopping forms (hopping mice, jerboas, kangaroos) is particularly apt in grasslands since large distances may be traversed with few jumps. Several species of bird also conform to this rapidly moving life-form type; ostriches, rheas, emus, bustards are all true savannah forms. Such animals often coexist as mixed herds, for example, in East Africa where herds of zebras, gnus, ostriches and antelopes occur. Carrion-eating vultures and high-flying prehensile species (eagles, kites) are conspicuous birds of this bioregion; and amongst the insects the termites, whose nests often give the landscape a characteristic appearance, deserve particular mention.

In contrast to the situation prevailing in the tropical rain-

* *Translator's note:* At least in Australia these are very much smaller than is desirable, and their area is decreasing at an alarming rate each year.

forest, the second biocoenotic principle applies here: the extreme habitat conditions cause animal populations to be poor in terms of species, and rich in terms of individual members. No other terrestrial ecosystems have such huge herds with millions of animals all of the same species: the North American bison, the passenger pigeon, African antelopes and – one of the oldest plagues of mankind – locust swarms.

As a climatic and vegetational zone, steppes are scarcely endangered by human interference unless radical changes result from burning or irrigation. Indeed, the clearing of dry forest areas by burning helps grassland to spread even further, as does the creation of fields, and, in suburban areas, of sports-grounds, etc. The original steppe fauna, however, has almost gone from many areas. Game mammals in particular, the bison and aurochs, the wolf and large cats, have almost completely disappeared, as have many bird species. Little hope exists for any of them because conservation areas (national parks) must be extremely large in order to provide them with natural living conditions. While such parks as Serengeti may be successful in Africa, it is already too late for the wild horse and bison of Europe and Asia.

Deserts

High temperatures and a lack of rain in the most extreme terrestrial bioregion, the desert, leads to an absence of vegetation for many months or sometimes years. The wind keeps the unvegetated sandy terrain in motion and thus constitutes another hostile environmental factor. Only nocturnal dew guarantees some moisture, and consequently active animal life is largely confined to the more favourable hours of darkness. Many desert animals, especially small rodents, live in caves or burrows during the day. Thus the burrowing life-form type occurs in addition to the hopping and running life-form types.

A spectrum of decreasing soil moisture and decreasing numbers of colonisers extends from the dry steppes to the semi-desert and from sandy deserts to stony and rocky ones. Nevertheless, plants and animals which are adapted to the special conditions exist in all these sorts of terrain. They only occur, however, in small numbers because of the low productivity of the ecosystem. Even in absolutely dry, mobile sand dunes of the African desert, where not the slightest amount of moisture is demonstrable throughout the year, there are subterranean insect communities (especially of the beetle family Tenebrionidae) as well as single rodents living off them. BRINCK (1957) has shown that these insects (as do other insects, for example the clothes moth, see p. 7) survive entirely without drinking, for they obtain water from their food by molecular transformation. This food is the organic dust swept together by wind action and found in the dunes.

So far, the great desert areas in the centre of continents have scarcely been exposed to human damage, and the same may also be said therefore with regard to their small animal fauna. This bioregion, at least then, poor as it may be, will be able to persist into the future.

Littoral Areas

Wherever marine or inland waters lie adjacent to a terrestrial bioregion, coastal areas, marshes, swamps and so on form a transitional region between the terrestrial and aquatic environment. The proximity of water, the resultant high humidity and the occurrence of occasional floods characterise these habitats. Their fauna is composed of descendants from both of the neighbouring regions as well as several specialised forms. Characteristic bird life includes waders, water rails and sandpipers, in addition to those which live in reedy areas and on mud flats and littoral sand. A diverse fauna of small animals develops in the habitat provided by the detritus deposited along the margins of the sea and inland water bodies.

This bioregion includes tropical mangrove swamps, which are forests of *Rhizophora* and *Avicennia*. These trees grow in suitable muddy places in the tidal zones of tropical coastal areas; their foliage projects above the water at high tide, whilst their prop roots are exposed at ebb-tide. The fauna is not diverse but it is well adapted to the special conditions, and includes the temporarily terrestrial mud-skipper (fish of the genus *Periophthalmus*) and the fiddler crab (*Uca*).

Sea

The ocean, with its enormous area of salt water and considerable uniformity, stands in complete contrast to the terrestrial bioregions. In its zone of open water, the *pelagial,* live the *plankton* (mostly small plants and animals in suspension) and the *nekton* (free-swimming animals: fish, cephalopods, jelly-fish). Plant production takes place in the illuminated zone (*eupelagial*),

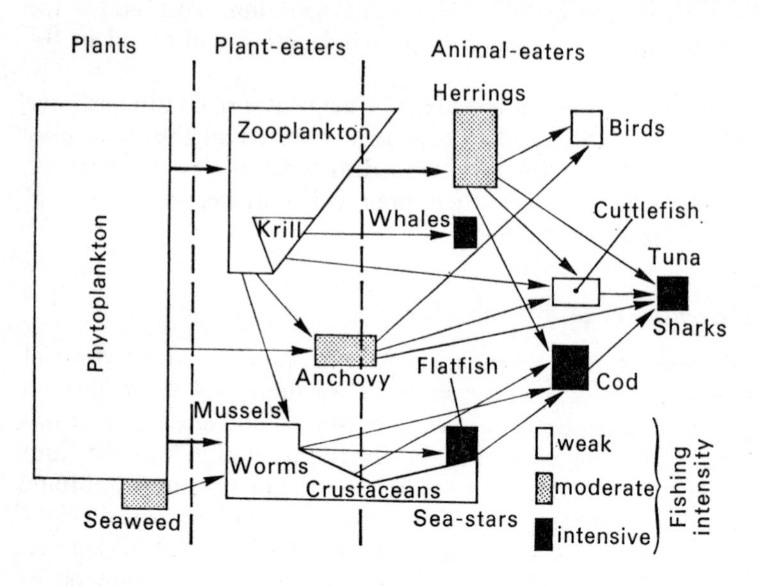

Figure 11 Trophic relationships in the sea (after Hempel, 1969)

while in the deeper zones (*bathypelagial*) only consumers are found, that is predators and animals which live off the plankton. In many oceanic areas, marine currents bring enriched deeper waters into the eupelagial so that greater production is established in such regions; increased fish production follows. Other oceanic areas are poor in nutrients and have small amounts of plankton and nekton.

The sea bottom constitutes the *benthal* habitat, and its population is referred to as the *benthos.* In shallow seas it comprises the phytobenthos (algae and sea-weeds) and the zoobenthos, the latter including both sessile animals (sponges, coelenterates, sea-feathers, tubicolous polychaetes, barnacles) and mobile ones as well (crustaceans, mussels, snails, various sorts of annelid, echinoderms). The zoological relationships of the deep-sea fauna (*abyssal*) are only imperfectly understood at present, although it is known that even in the deepest parts of the sea there is some oxygen, so that a benthic fauna can exist. However, this fauna is impoverished in both species diversity and, because of the low productivity, in individuals also.

On coral reefs, by contrast, there is a maximal development of living organisms. These reefs occur in all tropical seas and are formed by the deposition of large amounts of material derived from the chalk skeletons of reef-building animals and plants (especially of corals and bryozoans). In optimal light and thermal conditions, they represent the most dense and varied collections of marine fauna known because they also serve as a substrate for numerous benthic species and as a feeding area and refuge for many nektonic species (coral reef fish). Huge regions of the north-east coast of Australia are covered by a barrier reef, and in the Pacific the activity of reef-building organisms has given rise to atolls encircling slowly sinking volcanic elevations.

Inland Waters

Compared with the relatively uniform marine ecosystem, inland waters (that is, accumulations of water in terrestrial bioregions) show large variations in their physical and chemical factors, and very diverse faunas have therefore developed in them.

Inland saline waters (salt marshes, salines, salt springs) are inhabited by a fauna of euryhaline to stenohaline organisms as determined by the salinity. In accordance with the second biocoenotic principle, most populations in such waters are poor in species but have very large numbers of individuals. Moreover, insects (notably *Ephydra*), which are almost completely absent from the marine fauna, play a significant role in the fauna of saline waters. All other inland waters are fresh and contain only slight traces of dissolved salts so that they have a lower osmotic pressure than that of the body-fluids of freshwater organisms (whose body fluids are therefore *hypertonic* to the medium). Limnic organisms (freshwater inhabitants) are adapted to this physiologically unfavourable situation; they include many ciliates, several sponges, worms, insects, molluscs, crustaceans and freshwater fish.

Groundwater

The groundwater (*stygon*) comprises the enormous subterranean body of water which is exposed in caves (*troglon*) and springs (*crenon*) and which connects directly with rivers and lakes. As an ecosystem the stygon is very constant (*eustatic*) for neither oxygen concentrations nor temperature undergo seasonal fluctuation

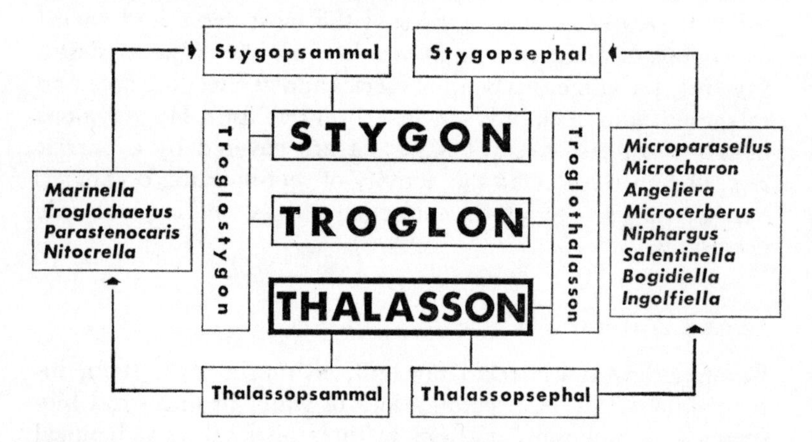

Figure 12 Groundwater ecosystems and their linkage through the immigration of colonisers (after Husmann, 1966)

and light is permanently absent. Temperatures always correspond to the mean annual isotherm of the region, and in temperate regions therefore always lie above freezing point. As a result, several very old faunal elements (Tertiary relicts) persist in this biotone: Tertiary crustaceans and water-mites. The scheme shown in figure 12 indicates the nature of the relationships between groundwater ecosystems; as shown, many characteristic animals of the fauna, for example *Niphargus*, originated from the sea (*thalasson*) and penetrated via coastal groundwaters into caves and inland groundwaters in the fine-pored sand (*psammal*) or coarse-grained gravel (*psephal*) of the subsurface.

Running Waters

Springs form from groundwaters that reach the surface, and from them arise flowing surface waters (brooks, streams). Whereas spring temperatures remain at about the same value throughout the year, those of flowing surface waters, being exposed to external temperatures, are more variable. Atmospheric oxygen is absorbed according to gradient and current speed. The benthic rheophilous fauna is well adapted to the special environmental conditions of this habitat and shows such specialisations as body flattening, suction discs, attachment devices, adhesive secretions and many others in order to resist being washed away by the current. A highly specialised benthic biocoenosis lives in the upper reaches of rivers (*rhithron*) where higher oxygen concentrations, lower temperature fluctuations and strong currents prevail. This biocoenosis shows extremely well those morphological and physiological specialisations which are adaptive responses to the flowing milieu, and it includes mayfly and stonefly larvae, freshwater limpets, amphipods, some caddis larvae and some chironomid larvae. Winter spawners preponderate amongst the fish (trout, char, grayling). In the lower reaches (*potamon*), temperature fluctuations are greater, and oxygen concentrations are more variable and lower. In addition, the current near the bottom is usually not strong. Eurythermous species tolerant of oxygen deficits live here (meso-eury-oxybiont species, see p. 15), and for the most part such species are also to be found in standing waters: dragonfly larvae, water beetles (Dytiscidae, Hydrophilidae), warm-adapted caddis flies (Limnophilidae, Atriplectidae), water lice (*Asellus*) and numerous water mites. The fish always spawn in summer (carp, pike and perch-like forms). The biocoenoses of the rhithron and potamon regularly occur throughout the world in relation to latitude and altitude as isocoenoses (figure 13). The only exceptions are localities where

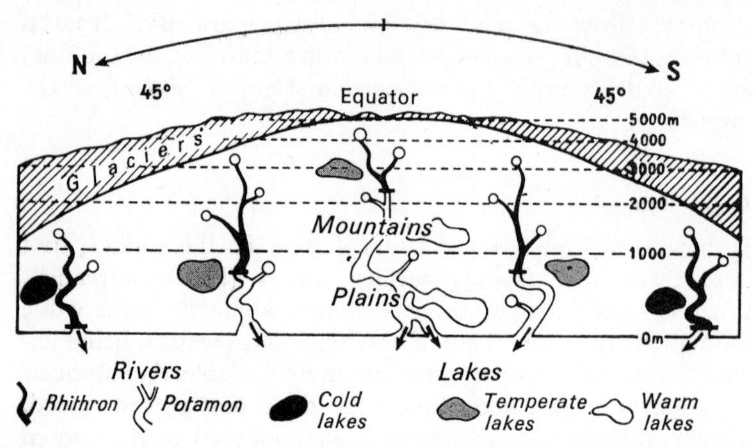

Figure 13 Regional distribution of inland waters (after ILLIES, 1961)

the extension of the rhithron biocoenosis is impeded, as for example on islands and isolated mountain chains (see p. 49).

Lakes

Exposed standing fresh waters form the ecosystem known as the *limnion* (lakes, ponds, pools). One of their most significant physical features is thermal stratification, which is brought about by solar heating and the anomalous fact that water has its highest specific gravity at 4°C. On the basis of their pattern of thermal stratification, FOREL distinguished three different lake types (figure 13). Cold lakes of the polar regions have their warmest water in the lower layers, while the upper layers are warmed during only a few months of summer and usually remain cold and ice-covered for many months. Warm lakes of the tropics and subtropics never freeze and throughout the year the bottom layers are coldest. And lakes in temperate latitudes change during the course of the year between both extremes, the change from a warm summer lake to a cold winter one taking place in autumn and the reverse change in spring. At these changeover times (the overturn periods) the lakes circulate fully, but during the intervening periods thermal stratification develops, especially during the period of summer stagnation. In the warm surface waters, a thermally homogeneous layer develops because of rapid circulation caused by wind action; below this layer (*epilimnion*) there is a 'discontinuity' layer (*metalimnion*) in which the temperature rapidly falls to a lower stable value in the deep layers (*hypolimnion*). The curve of oxygen concentration is the inverse of this one. The epilimnion is the productive stratum, whereas the hypolimnion, as determined by available oxygen, is the stratum of consumption and mineralisation. When oxygen is absent, a layer of mud is formed which is populated by only a few animals of the oligoeury-oxybiont type (*Chironomus, Tubifex*). The bottoms of tropical, stable stratified lakes can remain permanently without oxygen and animals (for example, Lake Toba in Sumatra). The epilimnion of lakes is dominated by a diverse limnoplankton (copepods, phyllopods, rotifers and protozoans among zooplanktonic forms, and diatoms, cyanophyceans and bacteria among phytoplanktonic ones) which constitutes the food base for the nekton (fish). Whereas the fish fauna shows strong regional differences, the zooplankton is relatively uniform and ubiquitous, with many cosmopolitan forms. Because of the absence of light, green phytoplankton does not survive in the hypolimnion, and it is here that bacteria are most abundant and where they undertake decompositional activities. The bottoms of standing water bodies and subsurface littoral zones are populated by the *benthos* which chiefly comprises insect larvae, oligochaetes, ostracods and aquatic snails.

According to the sort of balance that occurs in organic materials within lakes, two types of lake are usually distinguished, *eutrophic* and *oligotrophic* lakes. *Eutrophic* lakes have epilimnia with high production, and hypolimnia in which this can be only partly mineralised so that the excess production is deposited in annual strata on the bottom mud under conditions of deoxygenation. Most shallow lakes of the temperate latitudes are of this sort. In *oligotrophic* lakes, on the other hand, all production in the illuminated zone is consumed in the hypolimnion without there being any residual amount. Oxygen concentration remains high even at the bottom of the lake, and thus the fauna of the deep regions is abundant. All deep alpine lakes are examples of this type of lake.

Bogs

Quite different physico-chemical conditions prevail in bogs, and throughout the world result in a characteristic and uniformly represented ecosystem. They appear as the terminal phase in the filling-in process of those lakes having greater production than remineralisation. Above the mud and marshy layers, the development to flat and then raised bogs is led by the colonisation of peatmoss (*Sphagnum*) (figure 14). Strongly acidic water

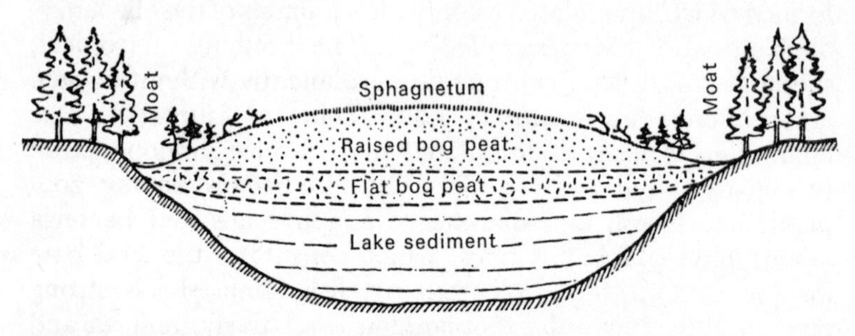

Figure 14 Cross-section through a raised bog, together with an indication of its history (after Ruttner, 1940)

with a high concentration of humus materials and a low concentration of minerals characterises bogs (*dystrophic* lakes). This biotope is colonised by a biocoenosis which is adapted to the prevailing chemical conditions, and the biocoenosis itself is derived from the vegetation of raised bogs and includes a few *tyrphobionts* (bog inhabitants) amongst the fauna (for example, several specialised insects).

PART II

Animals and Their Distribution

(Historical Animal Geography)

THE FAUNISTIC AND GEOLOGICAL BASIS

The task of historical animal geography is to establish the distribution of animal species and groups in geographically defined areas (chorology, faunistics) and then to explain it (causal zoogeography, distributional history). Faunistics, which naturally began first, is thus purely an accumulation of factual data and as such descriptive natural history in the old sense. It began well before mankind's scientific endeavours, for the knowledge of where certain sorts of animals lived and should be sought was one of the crafts of early human hunting and collecting communities. Even when scientific biology first began, with ARISTOTLE, faunistics was quite distinct and has continued so during the history of science. Information on where a species occurs, that is to say its restricted area of distribution, or 'habitat' according to LINNAEUS, is additional to and different from our knowledge of species characteristics and names. Zoogeographical knowledge in the sense of habitat knowledge exists in medieval documents of various sorts, but the idea of asking why a certain distribution occurred was pre-empted because, according to Genesis, the world had remained unchanged since its creation.

The faunal differences between single continents became especially obvious after the discovery of America, and BUFFON (1750) clearly perceived that the faunas of the Old and New Worlds were entirely different. The great scientific expeditions of the nineteenth century provided further details concerning

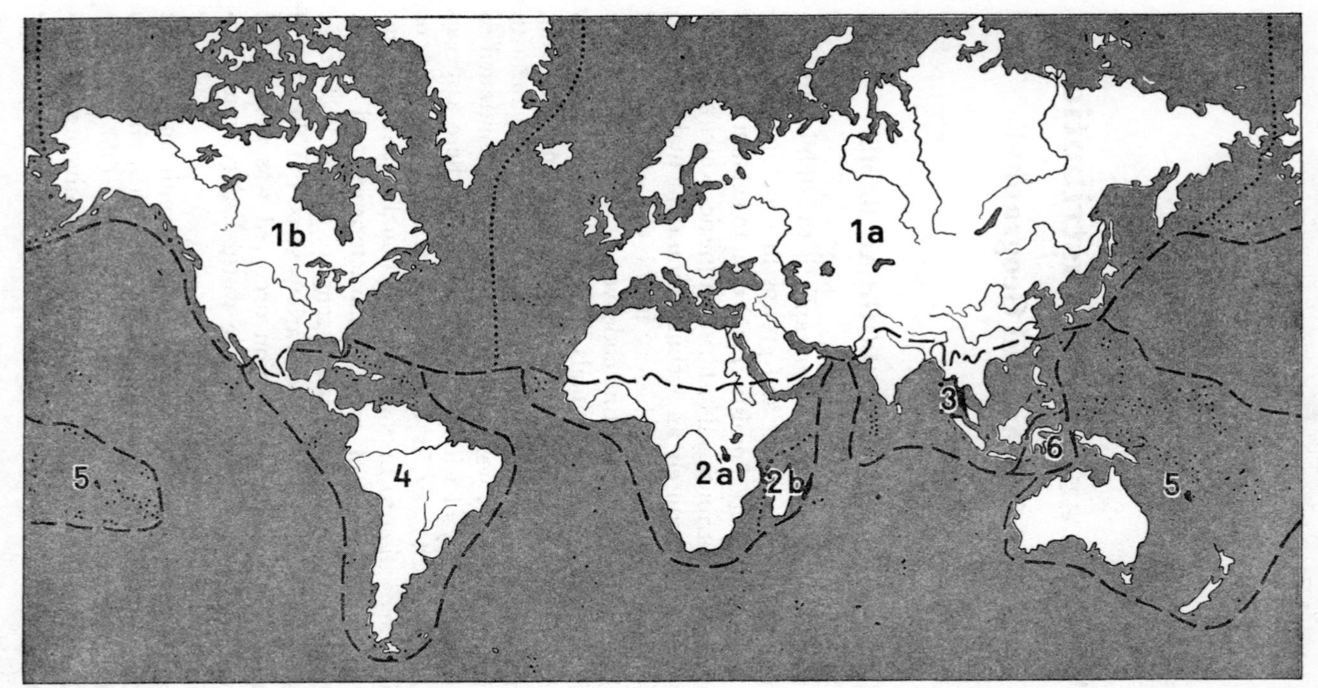

Figure 15 Faunal regions of the world (after De Lattin, 1967): 1a. Palaearctic; 1b. Nearctic; 2a. Ethiopian; 2b. Madagascan; 3. Oriental; 4. Neotropical; 5. Notogaea; 6. Wallacea

the world's zoogeographical regions, and they were demarcated with increasing precision by SCLATER (1858), HUXLEY (1868) and WALLACE (1878) (figure 15).

At first, animal distributions were explained on the basis of the account of Noah's ark given in Genesis: all species were regarded as having reached their present areas of distribution by spread from a central point, the ark. However, some modification of this account was proposed even as early as about AD 400, for AUGUSTINUS postulated individual creations after the great flood of the Bible in order to explain the presence of animals on islands. Similarly, two hundred years later, the Irish monk PSEUDOAUGUSTINUS suggested that the mammal fauna of his native island had arrived by immigration across a former land-bridge. With the discovery of America, the problem again became topical and the question of immigration across land-bridges was reconsidered in the sixteenth and seventeenth centuries (ZARATE, ALOSTA, KIRCHNER), while the ideas of AUGUSTINUS concerning separate and new creations that had been denied by orthodox theologians were reiterated by PARACELSUS and discussed for the last time by CUVIER in the nineteenth century. Scientific insight into causal zoogeography – in which both immigration and local speciation play a role – first began with the rise of evolutionary theory (LAMARCK, DARWIN) in the nineteenth century. (The independent development of finches on the Galapagos Islands provided one of the essential arguments for Darwin in conceiving his theory of evolution.)

For the historical analysis of faunal distributions, palaeogeographic statements are of great significance; if, for example, they provide proof of the former existence of land-bridges, then they offer zoologists the means of faunistic interpretation. However, it should be noted that the former existence of many land-bridges has been deduced from areas of disjunct faunal distribution so that the dangers of circular argument are great. The geological theory of permanence rejects every speculation of this sort and maintains that the present outlines and positions of the continents and of the ocean bottom have remained unchanged (WALLACE, 1876). According to this theory, all biogeographical phenomena can be wholly explained by migration across connecting pathways that still exist. An explanation of this sort is indeed possible for the more recent animal groups such as the mammals, and this view of zoogeography therefore still has some modern adherents (DARLINGTON, 1957). However, since geological scientists themselves have now abandoned the theory of permanence, a discussion of its biogeographical arguments is unnecessary.

In opposition to the theory of permanence, former land-bridges between now isolated areas have been increasingly invoked by biogeographers since the nineteenth century to explain present distributions. Thus FORBES (1864) regarded the colonisation of the British Isles as having taken place from the continent by a land connection across the English Channel; HOOKER (1847) postulated a trans-oceanic land-bridge between South America and Australia in order to explain their floristic similarities; and RÜTIMEYER (1867), HUXLEY (1870), SCLATER (1874) and VON IHERING constructed still further trans-oceanic land-bridges. Moreover, palaeontologists claim that there was a coherent southern hemisphere land-mass (Gondwanaland) during the Permo-Carboniferous period, and that the remains of this are the modern southern continents and peninsular India. However, as the various land-bridges and oceanic transgressions needed in biogeographical explanations are partly contradictory, poorly established and based on unconvincing geological concepts, land-bridge theories are insufficient when offered as explanations in causal zoogeography.

WEGENER'S (1912) theory of continental displacement (*epeirophoresis*) broke into this discussion and was a decisive turning point. According to this theory, biogeographically argued connections between separate areas were no longer to be regarded as synonymous with trans-oceanic land-bridges; in former times such areas had been in *direct* contact with each other. Several biogeographers (MICHAELSEN, IRMSCHER, RENSCH, JEANNEL) soon recognised that this theory offered a plausible explanation for the faunistic relationships between continents, and their consent became a strong argument for the theory. The geophysical basis of WEGENER'S theory, however, was rejected by most geologists, and many biologists were also not at first convinced (nearly always because their special group was unsuited to a discussion of it). As a result, the theory of continental displacement remained a debatable matter and the arguments for and against it finally became more or less equal and not so vehement. Then, after 'two decades of silence' (WUNDERLICH, 1964), the theory received new support, and now palaeomagnetic and geographical data (for example, deep sea movements) supported by stratigraphical (BEURLEN), geological (R. MAACK) and biological (BRUNDIN) evidence show that WEGENER's theory both in its fundamentals and in many details is certainly correct. According to the most recent geological information, the continents have been wandering since the Mesozoic and are derived from the disintegration of the original Palaeozoic continents of Gondwana and Laurasia. The continental drifting movements first involved the Pacific landmass and led to the formation of the circum-Pacific arc and at the same time of a latitudinal girdle of young, folded mountains

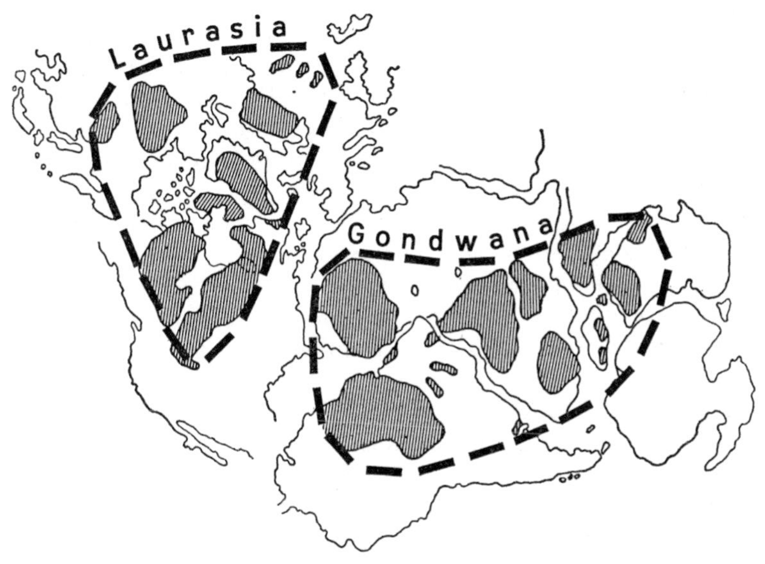

Figure 16 Reconstruction of the geographical arrangement prior to continental drift in which the continental nuclei, more than 1700 million years old, are grouped in two regions. These ancestral continents first broke up in the Mesozoic, and then drifted to their present positions (after HURLEY AND RAND, 1969)

(Atlas, Alps, Balkans, Himalayas, Sunda). Prior to these events, the primeval continents – of an age in excess of 4000 million years – had persisted unmoved over extremely long geological epochs and had been enlarged only by the deposition of more recent formations. This situation (figure 16) still existed in the Palaeozoic, the period when most modern animal groups originated, and the fragments which subsequently drifted apart were first released only some 200 million years ago (Mesozoic). All the land-masses involved in the continental displacements are inhabited by representatives of all modern animal groups and any historic or causal explanation of this distribution must involve the theory of continental drift.

BASIC PRINCIPLES OF CAUSAL ZOOGEOGRAPHY

Zoogeography is an empirical science which above all needs to collect and coordinate an abundance of factual data. Causal zoogeography as the intellectual mastery ('explanation') of these facts involves a process of mutual illumination between different subject areas (phylogenetics, palaeogeography, biological systematics). As the process remains incomplete and the theoretical framework has been derived only slowly, causal zoogeography cannot yet offer unequivocally formulated laws to the same extent as can other natural sciences. Nevertheless, some principles are apparent which may possibly provide the theoretical skeleton. They are formulated here for the first time.

1. *Principle:* The spread of some animal species can take place quickly and surmount serious barriers (marine barriers, mountains) and involves active migration or passive transport. Example: Krakatoa, a volcanic island some 40 km west of Java, experienced a tremendous eruption on the 26 August 1883 which destroyed most of the island and covered the remainder with layers of lava and ash to a depth of metres. All life was annihilated. A natural example of the new colonisation (re-colonisation) of a biotope took place in the period which followed. Botanical records of the recolonisation show that after three years 27 higher plant species were present, after 14 years,

62 plant species, and after 23 years, 114 species (the normal vegetation was then almost restored). The zoological analysis yielded: after six years, 40 species of arthropods and 1 reptile; after 25 years, 240 species of arthropods, 4 land snails, 2 reptile species, 16 species of breeding birds; and after 40 years, about 500 species of arthropods, 7 land snail species, 3 reptile species (1 snake, 2 lizards), 26 species of breeding birds and 3 species of mammal (2 bat species and the oriental rat). All these species reached the island by wind or marine currents or by their own powers of flight. The nearest island to Krakatoa is Sibesia, some 18.5 km distant, and the species involved must therefore have been carried at least 18.5 km by active or passive transport (in drift material). The original vegetation and fauna of this tropical island was restored at the latest after only 50 years – a minute time span in geological terms.

2. *Principle:* Slight obstructions (marine inlets, climatically unsuitable stretches) can prevent the spread of some animal species for long periods if there are no possibilities of transport and the innate tendency to spread is slight.
Example: In the fauna of the springs and streams of the Baumberge, some isolated hills near Münster (Germany), about fifty species of oligostenothermal benthic organisms are conspicuously absent, yet all of these occur in streams of the nearby highlands (for example, *Planaria alpina*, figure 1). The distance to the highlands (the western edge of the Teutoburg forest) is only about 30 km, and the Baumberge hills have been free of ice for more than 60 000 years. Thus, immigrants to the Baumberge streams after these became free of ice needed only to overcome the short (30 km) stretch of lowlands, and there were possibilities of transport by wind or water birds. In accordance with principle 1, most of the immigrants presumably overcame the barrier quickly, but the continued absence of some 50 species indicates that many benthic species may remain in a given biotope without ever leaving it even over long periods of time. It may also be noted at this point that anatomical or behavioural features which directly decrease the possibilities of being spread may develop, for example, the inability of many insects of mountain streams to fly.

3. *Principle:* Animal species that were widespread originally can be forced into one or more relictual areas (refuges) by environmental events (climatic changes, tectonic movements).
Example: In the western Pyrenees, the only locality for snails (Clausiliidae) of the genus *Laminifera* is the summit of La Rhune. However, Tertiary (fossil) representatives of this genus are known from several localities in central Europe. While central European populations were destroyed during the glacial period, the population on the summit of La Rhune obviously survived. (Mountain summits which remained free of ice during the glacial period and which played the role of refuges for animals are called 'nunataks' after an example in Iceland.) Recolonisation after the ice melted did not take place because the chances of distribution were slight and because competition from Clausiliidae established on the other mountain summits opposed a reimmigration of *Laminifera*.

4. *Principle:* Environmental events can cause the dispersion of animal species into widely scattered areas of distribution, and into areas of distribution where previously the same life-form type was unrepresented or represented by a different animal.
Example: In the boreo-alpine disjunction, as is shown for example by the ringed thrush (*Turdus torquatus*) (see figure 17), there are several areas of distribution that are quite isolated from each other. Some of these areas are in the boreal region, while others are in the high mountainous parts of central Europe and Asia Minor. Since the ringed thrush prefers cold climates and is not now encountered in the intervening warmer lowlands, its immigration into those areas where it presently occurs must have taken place at a climatically more favourable time. (The possibility of a contemporaneous 'origin' in several localities for a species is on principle rejected in biology as it contradicts the basic tenets of evolution.) During the Quaternary, the species obviously had a distribution in a boreal centre, perhaps one comprising the Alps and Balkans. From this it spread outwards during the glacial period when glaciers increased in extent and when the tundra had a more extensive border, encircled the

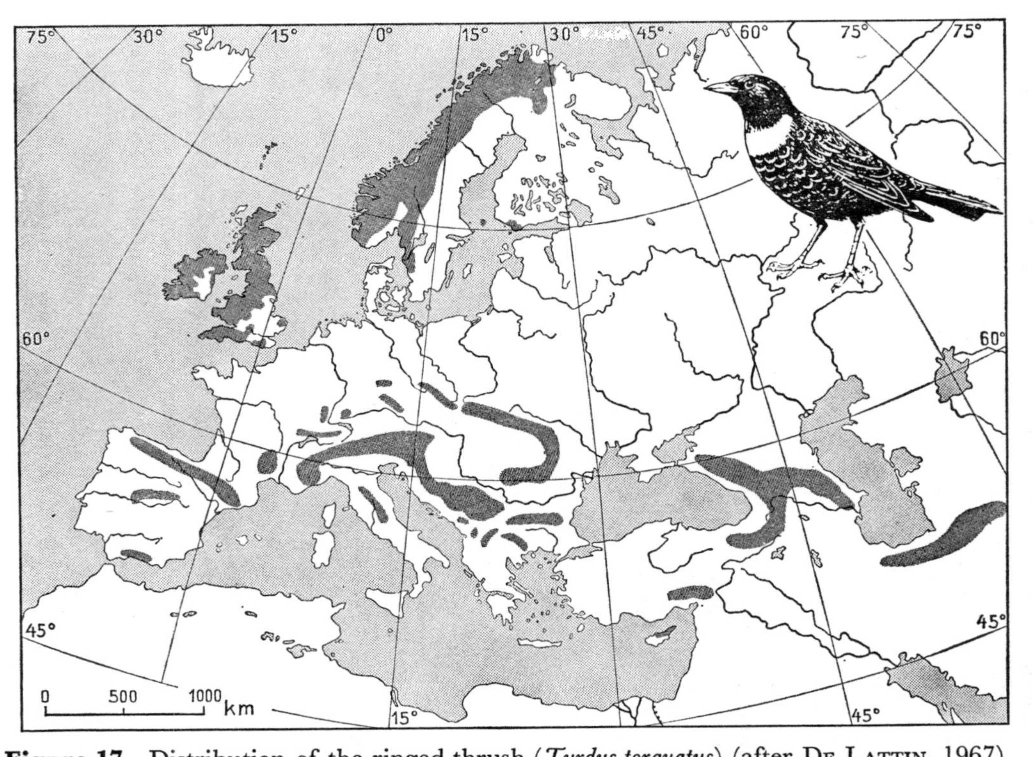

Figure 17 Distribution of the ringed thrush (*Turdus torquatus*) (after De Lattin, 1967).

mountains and extended to the adjacent lowlands. Thus, the boreal and oreal areas of tundra joined in the lowlands of northern Germany and a 'glacial mixed fauna' (THIENEMANN) formed. In the following interglacial period, the tundra receded northwards and southwards and took with it the cold-loving fauna. As a result the area of distribution became discontinuous and is still disjunct today. Its area of distribution is now the original boreal centre, other mountains of the wider region and parts of the boreal where the species was previously absent.

5. *Principle:* If a certain animal group currently shows a continuous distribution within an area, this area is usually the place where it originated.
Example: Darwin's finches (Geospizinae) represent a small subfamily of 14 species (in four genera) of finch-like birds, and they occur almost exclusively on the Galapagos Islands (one species also occurs on Cocos Island). This group of birds was discovered by DARWIN in 1835 during his world cruise and became for him the convincing evidence for the evolution of species: their presence could only be explained biologically by assuming that the recent species were derived from a common ancestor which had immigrated.

This principle has wide application: until proof to the contrary (see principle 6), it is usually assumed that an animal group originated where it is presently distributed. In many cases fossil evidence confirms this assumption.

6. *Principle:* Some groups of animals with a continuous (unfragmented) area of distribution have, nevertheless, immigrated into it and originated elsewhere.
Example: Giraffes are characteristic animals of the African fauna. As they are only found in this fauna at present, it appears that they originated in the Ethiopian region. However, this is not the case: in the Pliocene, giraffe-like animals occurred in south-eastern Europe, India, China and elsewhere, but not, on the contrary, in Africa. The group is thus extinct in the centre of its origin, and only a small off-shoot which migrated to Africa survived. It is now established in Africa as two species (giraffe and okapi).

THE FAUNAL REGIONS OF THE CONTINENTS

According to the degree of regional differentiation, the world's continents may be divided into various faunal regions (figure 15). Though these are very different in size, each is characterised by a specific fauna. Cosmopolitan forms, that is to say species which are common to all regions, represent only a small percentage of the fauna; they are almost exclusively protozoans, rotifers and other small forms which have resistant stages, spread easily, or are associated with human activity, frequently as parasites of man and his domestic animals. Conversely, endemics, that is to say species which are restricted to a region, form a characteristic component. Vertebrates and insects in particular are usually endemic to specific regions at the species and even higher taxonomic levels, and many of them are also endemic to considerably smaller areas. (The use of the word endemic, or of the term endemic form, can only meaningfully be used with reference to a distinct and specified area, for example, endemic to Europe, to Italy, to the Abruzzen district, etc.)

Of the six regions treated here, the first three are faunistically linked with each other in complex ways, and they have therefore since the work of Wallace (1868) and Huxley (1876) been considered together as *Arctogaea*, in contrast to *Neogaea* (Region 4) and *Notogaea* (Region 5). Arctogaea, with a certain degree of regularity, clearly shows the zoogeographical connection of the faunal regions involved, especially with regard to mammals; in the northern tundra, for example, the fauna is practically identical throughout Arctogaea, that is to say in both the Palaearctic and Nearctic (see p. 26). Southwards, however, there is a progressive differentiation of the Old and New World faunas, at first at the subspecific level: the reindeer (*Rangifer tarandus*), hare (*Lepus timidus*) and elk (*Alces alces*) provide examples. In the more temperate areas of the Nearctic, or as the case may be, the Palaearctic, the same genera occur but are frequently represented by different species (so-called 'vicarious species'), for example the marten (*Martes*), weasel (*Mustela*), wolverine (*Gulo*), lynx (*Lynx*), wolf (*Canis*), deer (*Cervus*), beaver (*Caster*), bison/aurochs (*Bison*) and horse (*Equus*). Still further

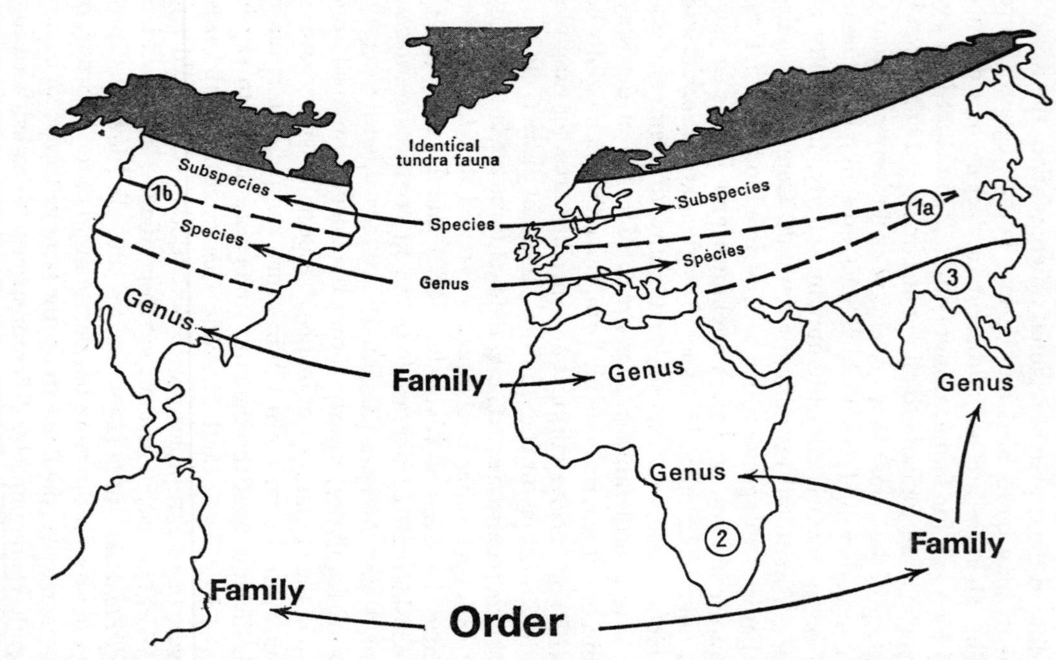

Figure 18 Faunal comparison within Arctogaea: increasing differentiation southwards

south, vicarious genera of the same family are found, for example moles (*Condylura* or *Talpa*) and prairie dogs (*Cynomys* or *Marmota*). This tendency for the progressive differentiation of the fauna southwards in Arctogaea, indicated schematically in figure 18, continues south even beyond the limits of the Holarctic so that even the rank of vicarious families is occasionally attained, as, for example, in the case of the New World apes (*Platyrrhina*) and Old World apes (*Catarrhina*) or, respectively, llamas and camels. The explanation of this conspicuous faunal differentiation may lie in the phenomena of continental drift and the Mesozoic separation of the continents, and in the consequent southwards spread of more recent animal groups such as the mammals.

Holarctica

The faunistically uniform and enormous land mass of the Holarctic comprises the Eurasian continent from its northern border to the Sahara and Himalayas (*Palaearctic*), and the North American subcontinent down to the Mexican state of Sonora (*Nearctic*). It occupies a large part of the northern hemisphere. The history of the present populations of this region, in the main, reflects the climatic variations of the Ice Age. During the period of maximal Pleistocene glaciation, when because of decreased sea-levels the Bering Straits did not separate Asia and North America, these continents were in terrestrial continuity over a wide area. As a result, the fauna and flora which had been forced south in front of glaciers from northern and central continental areas must have become thoroughly mixed, and eventually the existing faunal differences of Tertiary times disappeared and a uniform, mixed Holarctic fauna was formed. After the retreat of the glaciers, the loss of the terrestrial connection, and the recolonisation of newly exposed areas, fresh regional faunal differences developed in which the degree of systematic separation of the Nearctic or Palaearctic faunas reflected the duration of isolation. Since larger areas of the Palaearctic (central Siberia) remained unglaciated, these served as starting points for an early recolonisation of those areas newly freed from ice in Eurasia and – via Beringia – in western North

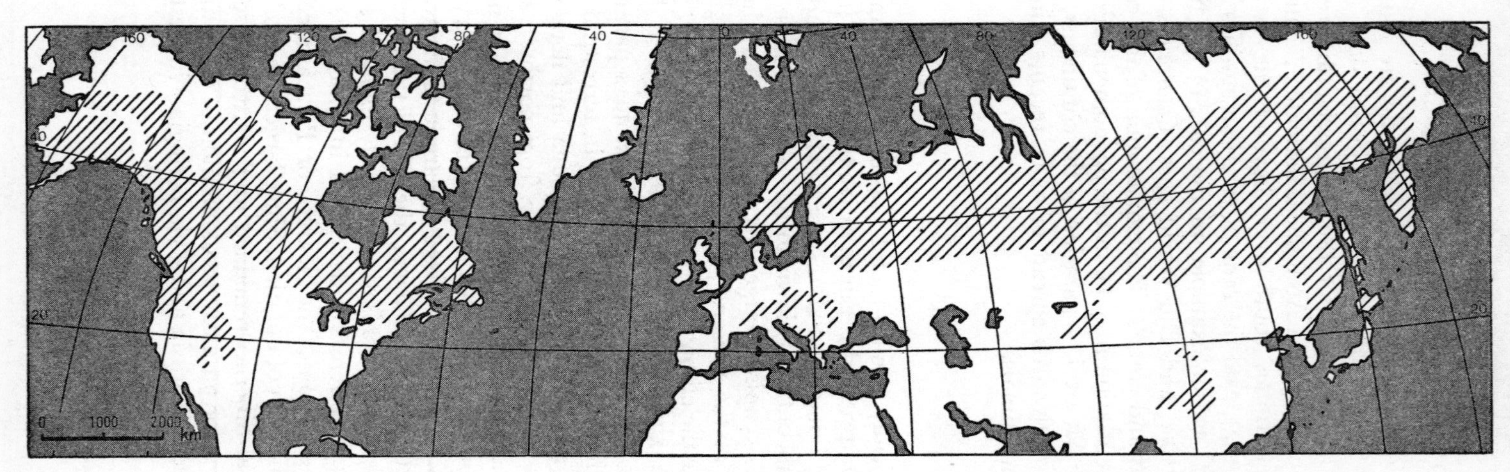

Figure 19 The distribution within the Holarctic of the three-toed woodpecker (after DE LATTIN, 1968)

Plate 1 Mixed herds in an African savannah

Herds of white-bearded gnus and small groups of zebras – often with springboks and ostriches – still form a characteristic feature of the East African landscape over wide areas. The thick, year-round vegetation of hard grasses of the open savannah offers an adequate food base for a diverse association of grazing forms. By a combination of warning and flight-reactions, the mixture of different species improves the chances of survival against various natural enemies, hyenas and lions

Plate 2 Australian savannah with termite mounds

The scattered earthen constructions of this insect characterise the landscape. Several million individuals can exist in each of the cement-hard structures, which are up to three metres high and extend deep into the ground. Food requirements (preponderantly wood) and active fighting (by special 'warriors') are responsible for the fixed minimal distances between mounds and their uniform spacing within the available area

Plate 3 Bird-inhabited rocks in Peru (pelicans on the rocky coast in front of St. Margarita)

Large numbers of pelicans and some other fish-eating coastal birds often use secure positions near water as nesting, sleeping or breeding places. Their phosphate and nitrate-rich faeces are deposited in such places in layers that are often metres thick, and which can later be mined and exported as mineral fertiliser

Plate 4 Coral reef off the New Caledonian coast

In shallow, tropical parts of the sea, corals and polyzoans form submerged banks by intensive growth of and constant deposition upon their chalk skeletons. These banks develop into extensive plateaux or atolls, and, following an elevation of the sea floor, may give rise to islands. Coral reefs represent one of the most densely populated of marine habitats and are characterised by a diverse and colourful community of worms, molluscs, crustaceans and fish (see p. 37)

Plate 5 Sheep and shepherd on Lüneburger Heide

The pine–heath landscape – with juniper and flourishing heath the very essence of a romantically beautiful natural area – originated in fact from the oak–birch forest that was climatically native to the area. The change was brought about by the constant browsing and overgrazing of a certain sort of heathland sheep (a northern type). The completely uneconomic shepherds are nowadays subsidised by the state in order to prevent the displacement of the heath by forest and to retain the familiar landscape

Plate 6 South polar penguins (*Pygoscelis antarctica*) in the Antarctic

The penguins are the only warm-blooded animals living on the unvegetated rocks and glacial areas of the Antarctic. They are incapable of flight but are adroit swimmers and divers, feeding exclusively on marine fish. They seek land to rest and breed, and then gregariously congregate in enormous flocks (see p. 86). In the hostile icy wastes they represent the last outposts of life

Plate 7 Flamingoes on Lake Nakuru in Kenya

Amongst littoral wading birds, the gracious, delicately pink-coloured flamingoes are the most conspicuous and permanent species determining the face of the landscape. They feed on small animals which they catch in the shallow marginal zones by standing in the water and using their beak as a filtering device. They are often found in huge flocks

Plate 8 Marine lizards on the rocky coast of the Galapagos Islands (see p. 75)

Sunbathing on the coastal rocks of the Galapagos Islands, the grey–green and approximately one metre long lizards of the genus *Amblyrhynchus* present a living picture from the Mesozoic. On these islands they have no enemies (predators), and feed on submerged marine algae which they graze by dexterous diving and swimming. Large groups of them congregate on warm, stony surfaces

America. Consequently, the faunistic similarities (for example, in many insects) between Europe and western North America are distinctly greater than are they between Europe and the geographically much closer region of eastern North America where, indeed, as a result of the much later glacial retreat, many of the genera are absent.

During the course of the climatically induced faunal displacements, disjunctions frequently developed in the cold-preferring arctic species (boreal or oreal), as described in principle 4 (see p. 50). Examples of this are given by the boreo-alpine disjunctions of Europe (figure 17), and similar distributional patterns are known from the Nearctic and the eastern Palaearctic (Japan–Korea). As the Pleistocene climatic deterioration progressed, the forest-inhabiting Tertiary fauna was forced far southwards and finally could survive only in the few remaining forested areas of eastern Asia, the Mediterranean region and the south-eastern Nearctic region. A series of species did not return to their original areas after the Quaternary increase in temperature, and now occur in greatly disjunct Palaearctic areas (figure 20).

The warm-preferring Tertiary fauna of the Holarctic was almost completely driven from this region in the Pleistocene, and was largely destroyed except for that part able to reach subtropical areas of the Ethiopian and the Oriental regions. Numerous fossil representatives of this Holarctic Tertiary fauna are known from Tertiary deposits and Baltic amber (Oligocene), but few of these species are still extant, and most belong to groups now occurring in the subtropics. Those Tertiary species which have survived do so as relicts; they persist only where climatic conditions are especially favourable, particularly in certain freshwater biotopes. Three such limnic biotopes are especially important:

The Groundwater. The special temperature conditions operative in groundwaters (see p. 38), and in particular those in thermal groundwaters, have enabled this biotope to become a refuge for several members of a Tertiary freshwater fauna otherwise extinct. *Thermosbaena mirabilis*, a crustacean from North African thermal waters (48°C), is the only extant representative of the subclass

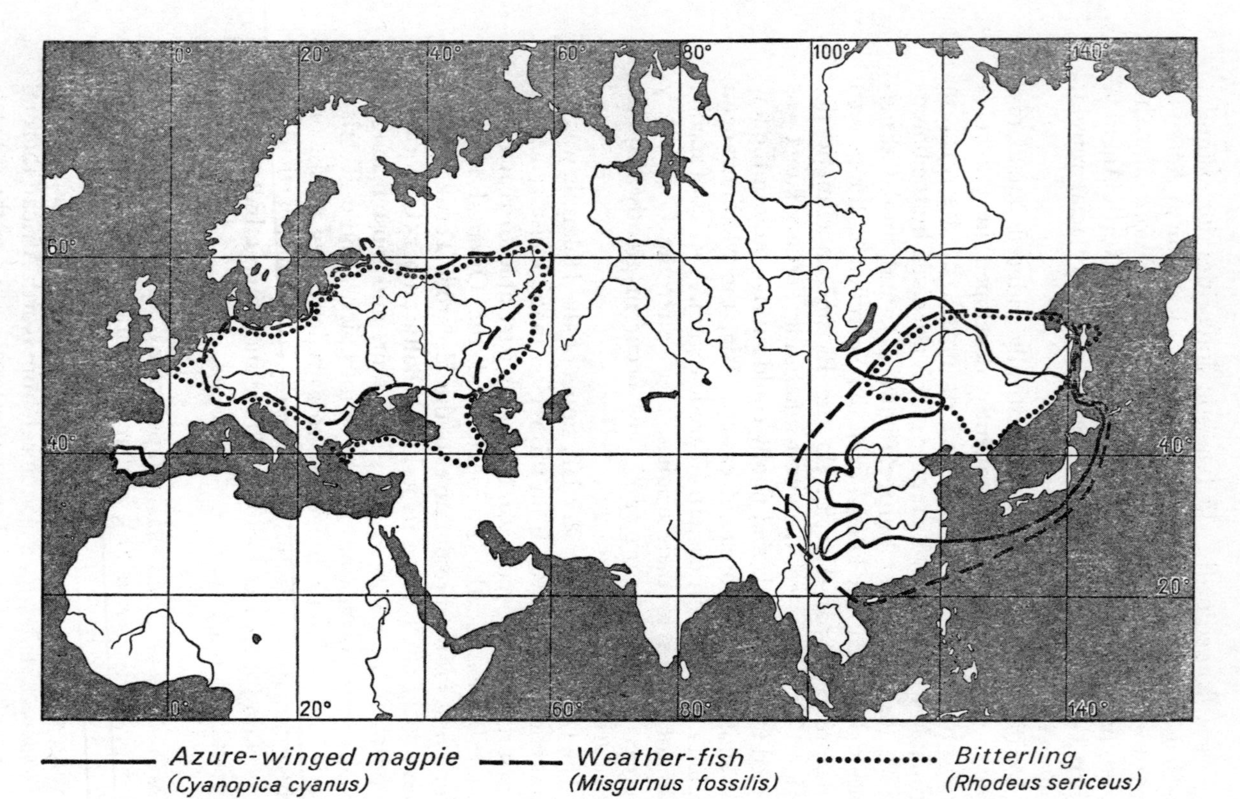

Figure 20 Examples of European–East Asiatic disjunctions (after RENSCH, 1950)

Pancarida. The snail, *Melanopus pereyssi*, which occurs in thermal waters (34°C) at Bischofsstadt (Hungary), belongs to a family elsewhere now occurring only in the tropics and subtropics. However, groundwaters of normal temperature also possess a crustacean fauna of distinctly primitive, old forms: Bathynellidae (Syncarida), *Parastenocaris* (Harpacticidae), *Ingolfiella* (Amphipoda), for example, are all obvious Tertiary relicts which have survived in groundwaters; they have assumed a slender and small body form adapted to the habitat and its narrow interstitial system. In larger accumulations of groundwater (lakes in caves), the cave salamander or Proteidae has survived (Nearctic, *Necturus*; Palaearctic, *Proteus*).

Ancient Lakes. Several Holarctic lakes distinguished by great depth and a certain geographical position have a history extending back to Tertiary times and have also remained completely free of ice during the Ice Age. A series of endemics is still preserved in their fauna and these must be regarded as direct descendants of a Tertiary fauna otherwise extinct in the Holarctic. Lake Ohrid and, though to less extent, the neighbouring, smaller Lake Prespa on the Greek–Albanian border represent such Tertiary refuges in Europe: endemics here include freshwater sponges (*Ohridospongia*), flatworms (*Neodendrocoelum*), 24 of a total of 26 snail species, numerous crustaceans and a fish genus (*Pachychilon*).

The relictual character of Lake Baikal is even more distinct; this lake is 70 million years old. A sponge family (Lubomirksidae), a snail family (Baikalidae), and even a fish family (Comephoridae) are endemic here as well as numerous genera and species from almost all other limnic families. The development of the crustacean order Amphipoda is particularly marked; there are 240 species in 34 genera endemic to the lake – which thus contains about a third of the world's amphipod fauna. The only living limnic seal (*Phoca sibirica*) is also found here, although only as a relict from the period of maximum glaciation.

Other ancient lakes of the Holarctic are the Lake of Tiberias (Sea of Galilee) in Israel, and Lake Tahoe in California; their fauna, however, is not as rich in relictual forms.

Warm-temperate Estuaries. In two places in the southern part of the Holarctic, an ancient fauna has persisted in the deltaic regions of large rivers draining south or east, namely the Mississippi and the Yangste Kiang. The fauna of the lower courses of these rivers provides, as Tertiary relicts, a last reflection, so to speak, of the former subtropical fauna. It includes crocodiles (*Alligator mississipiensis* or *A. chinensis*, respectively), giant salamanders (*Cryptobranchus* or *Megalobatrachus* – also known from fossils in Europe) and archaic fish of the sturgeon order (the genera *Polyodon* and *Scaphirrhynchus*, each with a single species).

By far the majority of Holarctic animals belong to families and orders which are also to be found outside the Holarctic; the evasive movements of the Tertiary fauna during the Quaternary have largely been responsible for this (in accordance with principle 4, see p. 50). Thus, many characteristic faunal elements of the Holarctic also occur in the Oriental or Ethiopian regions; among mammals they include in particular hedgehogs (Erinaceidae), hares (Lagomorpha), martens (Mustelidae), dog-like forms (Canidae), bears (Ursidae), horses (Equidae), cattle (Bovidae), deer (Cervidae) and pigs (Suidae). Any listing of Holarctic endemics, in contrast to the situation in other faunal regions, is essentially not one which is of animal groups particularly typical to the region, but one which is merely a selection of those families having (accidentally) no Recent representatives in regions other than the Holarctic. Examples are provided for mammals by the beaver, mole and chamois, for birds by divers, auklets, waxwings and finches, for amphibians by the Proteidae, salamanders and discoglossid frogs and for fish by the pike.

At the lower systematic rank of genus, numerous endemics frequently occur that have different species in the Nearctic and Palaearctic. Insect examples are given by the Apollo butterfly (*Parnassius*) and the large ground beetle (*Carabus*). However, Holarctic genera by no means always occur in Nearctic–Palaearctic species-pairs (as indicated schematically in figure 18), but often have numerous endemic species in substantially smaller areas. The genera just mentioned contain dozens of endemic species, with every large mountain chain often having one. (Because collectors have placed considerable value on these and have observed them intensely, there has been a marked differentiation of them into subspecies, races and varieties; in the most extreme situation, the inhabitants of single valleys have been given a special name.)

This detailed faunistic structure represents a measure of the climatic and geological history of the area involved, and in particular identifies centres of isolation (mountains, islands). Almost exclusively, it involves the systematic ranks of species and subspecies, and in principle it may be observed in all faunal regions. The phenomenon may be illustrated by examples from the heavily investigated freshwater fauna of Europe. Europe has been divided into 25 zoogeographical regions (see figure 21) (for details and rationale concerning the division see ILLIES, 1967). From the total area, some 12 015 limnic species of multicellular animal are known. Perhaps a sixth of these (1912) are endemic to single regions, with the Iberian peninsula providing 276 endemics, the Hellenic–Balkan area 272, Italy and associated islands 263, the Caucasus 225, the Dinaric west Balkan area 200, the Alps 115, the Carpathians 97, the Pyrennees 75 and the east Balkans 58 endemics. All other regions, particularly those in the north, are poor in endemics; in the lowlands there are perhaps only ten species per region. In this enumeration, the abundance of endemics in Lake Ohrid (see p. 59) plays a special role in the case of region 6, as does also the isolated mountain fauna of the island of Corsica for the inhabitants of running waters in region 3 (figure 21).

The numerical insignificance of endemics in terms of the total fauna of a region is illustrated by the freshwater fauna of the Alps (region 4 of figure 21). Altogether, 4146 species are reported from this region but only 115 of them are alpine endemics. Perhaps 100 species are boreo-alpine forms, that is occur in regions 4 and 20–22. Perhaps another 200 occur in regions 4 and 10 (endemic Carpathian-alpine forms). About 2000 species occur in highlands bordering the northern edge of the Alps (regions 4, 8, 9, 10), whilst others (progressive southern species of the edge of the glaciers) occur in the southern highlands of region 3 or in the north-western Balkans (region 5). Finally, there are about 1000 species native to the whole of

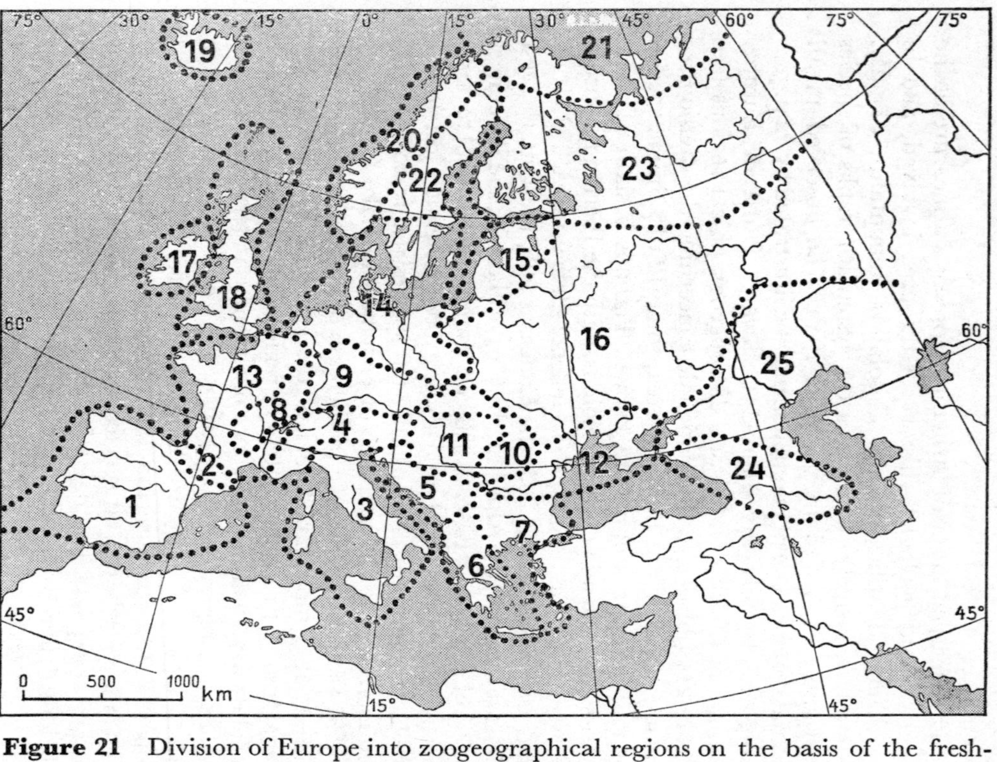

Figure 21 Division of Europe into zoogeographical regions on the basis of the freshwater fauna (after ILLIES, 1967).

Europe and extending in part as far east as at least the Urals (regions 16 and 23).

Ethiopian–Oriental Regions

Abutting southwards on to the great Eurasian continental block are two zoogeograpical regions, the Ethiopian and the Oriental. Although they lie adjacent to the Palaearctic well north of the equator, both also project far into the southern hemisphere. These regions have many common features, and because of this are treated together here to begin with. Both were covered during the Tertiary by tropical rainforest inhabited by groups of animals now found in such habitats throughout the world (including those in South America), for example, crocodiles, parrots, finfoots (Heliornithidae), pythons, geckos and caecilians (Gymnophiona).

During the Pleistocene climatic deterioration, those inhabitants of Palaearctic forests that preferred warm conditions advanced southwards into steppe areas. These, contemporaneously, were spreading and replacing the receding forest, so that ultimately rainforests became restricted to equatorial regions and the previous continuity of the forested regions became disrupted. The steppe areas themselves became isolated by the Persian highlands, and a complete separation of the Ethiopian and Oriental faunas resulted. Many of the families endemic to the two regions consequently developed unique genera in each region, as indicated in figure 18. The mammals provide many examples of such vicarious genera: for example the great apes (the chimpanzee and gorilla in the Ethiopian region; the orang utan and gibbon in the Oriental region) and, respectively, the monkeys (*Rhesus*, mandrills and baboons; leaf monkeys (*Presbytis*) or holy apes of India), lorises (galago; slender loris of India), elephants (*Loxodonta*; *Elephas*), rhinos (*Diceros* and *Ceratotherium*; *Rhinoceros* and *Didermocerus*), native cattle (the buffalo, *Synceros*; the zebu, *Bibos*), forest antelopes (the kudu, *Strepsiceros*; the Indian bluebuck, *Boselaphus*), pigs (the African water hog, *Potamochoerus*; the Oriental–Palaearctic *Sus*) and civets (*Civettictis*; *Viverra*). The unique mammalian order of scaly ant-eaters or pangolins (Pholidota) has a single genus, *Manis*, with four Ethiopian species and three Oriental ones. Finally, the aardvark (*Orycteropus*), now endemic to Africa, is known from fossils in India. Amongst bird families endemic to the Ethiopian–Oriental area may be mentioned the sunbirds (Nectariniidae), hornbills (Bucerotidae) and bulbuls (Pycnonotidae).

In several animal groups there was an exchange or an immigration of a few species. Thus, the chameleons are essentially of African origin but are also represented by several species in the Oriental region, a phenomenon that applies to the scaly ant-eaters (Pholidota) as well. Then there are several species that are common to both regions as the result of a later immigration; this is the case for the great cats in particular, that is the lion, tiger, panther and cheetah.

Ethiopian Region

Africa south of the Sahara corresponds to the zoogeographical region known as the Ethiopian. Those animal groups in this region which have never been discovered as Palaearctic fossils are to be considered as its archaic faunal element: amphisbaen lizards (Amphisbaenidae), caecilians (Gymnophiona), cichlid fishes (Cichlidae) and freshwater mussels of the family Mutelidae. Since all these groups occur in South America (Neotropis) as well, they have always provided a strong argument for the existence of the Palaeozoic continent, Gondwanaland (figure 16), and since hardly anyone now denies the existence of this continent, it follows that these Gondwanaland elements of the Ethiopian region represent Mesozoic relicts of the African fauna. Gondwanaland elements also occur in the arthropod fauna of the southern tip of Africa, particularly in many groups of insects and the Palaeozoic arthropod order containing *Peripatus* (Onychophora). Those relicts occurring in the freshwater fauna of ancient lakes probably stem from Tertiary times, and reflect a situation similar to that previously described for ancient lakes in the Holarctic (p. 59). Lake Tanganyika (depth, 1270 m) contains a series of such endemic forms, including about 30 genera of the fish family Cichlidae, a freshwater genus of herring and a freshwater jelly-fish.

During the Ice Age there was a massive immigration of Palaearctic groups and it is these which now characterise the

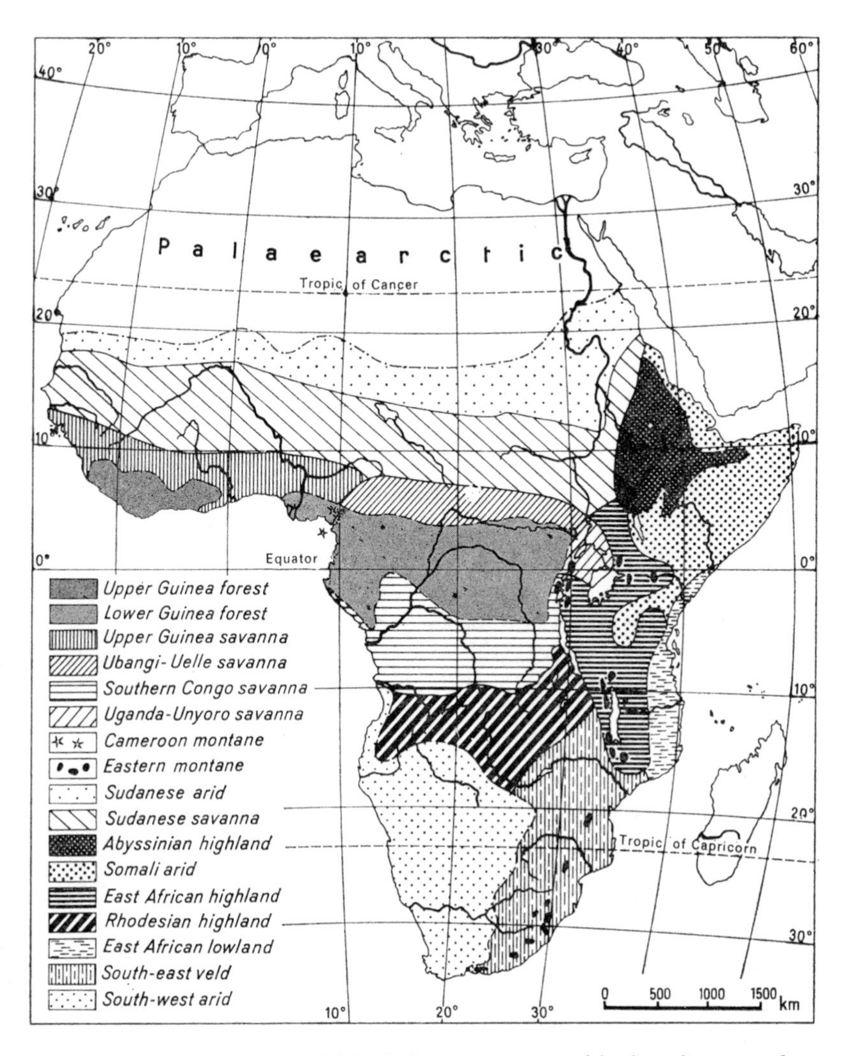

Figure 22 Division of Ethiopis into zoogeographical regions on the basis of bird distributions (after CHAPIN, 1923)

fauna, especially in the African savannahs (figure 22). Extant endemics amongst mammal families include: giraffes (see, however, p. 52, principle 6), hippopotamuses (Hippopotamidae), elephant shrews (Macroscelididae), otter shrews (Potamogalidae), golden moles (Chrysochloridae), scaly-tailed squirrels (Anomaluridae), aardvarks (Tubulidentata), African jumping hares (Pedetidae), maned rats (Lophiomyinae), African dormice (Graphiuridae), conies (Procaviidae), sable and roan antelopes (Hippotraginae), African mole rats (Bathyergidae) and the monkey-like lorises (Galaginae). On the other hand, the complete absence of deer (Cervidae) and bears (Ursidae) is notable. Several families of birds are also endemic to this region: for example, wood-shrikes (Prionopidae), colys (Coliidae), wood-hoopoes (Phoeniculidae), touracos (Musophagidae), guinea-fowls (Numididae), secretary birds (Sagittariidae), whale-billed storks (Balaenicipitidae), ostriches (Struthionidae) and hammerheads (Scopidae). Considering freshwater fish, the families of the birchir (Polypteroidei), snoutfish (Mormyridae), certain catfish (Malapteruridae) and the African lungfish (Protopteridae) are all endemic.

This unusually large number of endemic families, taken with the endemic genera of Ethiopian–Oriental families (see p. 63), gives the vertebrate fauna of the Ethiopian region a particularly individual note: the African savannahs with their mixed herds of antelopes, zebras, gnus and ostriches present a zoogeographically unique and unmistakeable picture.

The detailed faunistic structure of the region with regard to birds is indicated in figure 22. The equatorial jungle region of the Congo area which extends eastwards to the adjoining great lakes and the Abyssinian highlands (figure 22) acts as a strong faunal barrier and partitions the continent into two disjunct areas of savannah and steppe. Many genera of animals occurring in these areas have vicarious and endemic species, so that the northern and southern subtropical parts of Africa are clearly distinguishable. The oreal fauna of the highlands includes numerous endemics of species rank, but most of the cold-stenothermal animal groups are lacking even at the snow-line on Kilimanjaro; a genuine endemic fauna, such as occurs in the Alps, is absent.

Madagascar (Malagasy)

This zone includes the island of Madagascar itself and the island groups of the Comoros, Seychelles and Mascarenes. In spite of the proximity to Africa, the Madagascan fauna exhibits considerable individuality, as indicated, for example, by the fact that almost all Ethiopian mammals are absent from it; only the African water hog (*Potamochoerus*) has reached the island as a later immigrant from the African continent. Such Ethiopian and especially Oriental groups as do occur are represented by endemic genera; and the civets (Viverridae) are noteworthy as the only predators present. During the Mesozoic, Madagascar had a close connection with India to form a Gondwanaland nucleus (figure 16). SCLATER, in order to explain the close Madagascan–Oriental faunal relationships, believed that there must have been a land-bridge across the Indian Ocean and this bridge he thought was the hypothetical continent of 'Lemuria'. However, in the light of the theory of continental drift, the hypothesis is superfluous.

Amongst the endemic groups, the mammals are particularly conspicuous; they are represented by six families: three monkey-like families (Lemuridae, Indridae, Daubentoniidae), a bat family (Myzopodidae), the tenrec (Tenrecidae) and the Madagascar rats (Nesomyidae). With regard to the avifauna, the following are especially noteworthy: the elephant-bird (Aepyornithidae), which has become extinct only within historical time, mesites (Mesoenanatidae), asitys (Philepittidae), cuckoo-rollers (Leptosomatidae) and the Brachypteraciidae.

The Madagascan fauna is particularly characterised by the poverty of life-form types, as shown by the absence of many Arctogaean elements which are otherwise widely distributed: even-toed ungulates, horses, apes, carnivores (except for civets), hares and mice, as well as many bird families (finches, woodpeckers, cranes) and reptiles (lizards, guanas, agamids, venomous snakes).

Oriental Region

The eastern zoogeographical region referred to as the Oriental (figure 15) comprises tropical Asia with India, Indo-China, the

East Indies, the Philippines, Formosa and the southern part of China. As such, it represents, as it were, a southern extension of the Palaearctic, from which region, however, it is separated by the Tibetan highlands. Difficulties involved in delimiting it from the adjacent eastern region are discussed on p. 69 (figure 23). Its fauna is especially characterised by Ethiopian–Oriental elements (see p. 63), but a series of Palaearctic groups which immigrated during the Ice Age into the Oriental region and not the Ethiopian is also present. The mammals include bears (Ursidae), deer (Cervidae) and, in the Bovidae, sheep and chamois.

Endemic mammal groups that have developed include the moon rats (Echinosoricinae), tree shrews (Tupaiidae), colugos or flying lemurs (Dermoptera), cloud rats (Phloeomyinae), shrew-like rats (Rhynchomyinae), pandas (Ailuridae), spiny dormice (Platacanthomyinae), tarsiers (Tarsiidae) and gibbons (Hylobatiinae). Among birds there are only the leafbirds (Irenidae) and the crested swifts (Hemiprocnidae).

Although these groups confer a degree of a unity upon the Oriental fauna, it is still possible to subdivide the region into three relatively heterogeneous districts, namely, India, Indo-china and a China–eastern Himalayan district, the faunas of which are clearly distinguishable at the species level. Even in the Eocene, these three districts were entirely isolated from each other by marine incursions; moreover, during the Mesozoic, India was part of Gondwanaland and was closely connected with Madagascar. However, active tectonic movements appear to have largely destroyed the Gondwanaland fauna of India, and this fauna was then replaced by immigration and recolonisation from the Palaearctic. Thus the fossil mammal fauna of India displays more relationships with the Ethiopian fauna than with the extant Indian fauna. Oriental and Palaearctic elements are interspersed widely throughout the Chinese–Himalayan district so that their delimitation is problematical; in the main, it parallels the extension of the forests.

Details concerning the faunistic structure at the level of species and superspecies enable the demarcation of distinct subdivisions of the region. The Sunda area has been particularly well investigated in this respect (it has been a classical area for such zoogeographical investigations since the time of WALLACE, and in the twentieth century it has been investigated by WEBER, THIENEMANN and RENSCH). The work of MOOLENGRAF and WEBER (1919), for example, provides clear details of zoogeographical relationships within the Sunda islands; they found that the rivers of northern Sumatra and western Borneo possessed a very similar fish fauna (of 119 species present in the Kapuas River in west Borneo, 51 were also present in Sumatra), whereas in the rivers of eastern Borneo there are only very slight relationships with the south (in the Makaham River, only five of 75 species). The explanation for this phenomenon is to be found in the Pleistocene history of the Sunda arc: the depression of the sea-level during the Pleistocene united Borneo, Sumatra and Java with the Malayan peninsula to form 'Sundaland' (figure 23) in which the Kapuas River and Sumatran River were tributaries of a large river which flowed northwards into the China Sea, and whose course is still demonstrable as a channel on the sea-bed. Its autochthonous fish fauna survives as a relict in the contemporary Kapuas River. On the other hand, the Makaham River, already isolated at that time, flowed eastwards and thus developed its own endemics.

Wallacea

Because of its peculiar transitional characteristics, the East Indian–Australian area which adjoins the Oriental region eastwards has long been allocated the rank of a special zoogeographical region. Its contemporary fauna in any event does not permit these islands to be attached unequivocally to either the Oriental or Australian regions. The so-called Wallace's line is considered to be the western edge of the region, that is the limit of the extension of Australian animal groups (for example, in marsupials, the cuscus), while the eastern edge is Lydekker's line, the furthest penetration of Oriental forms (for example, the flying lizard, *Draco*). Weber's line represents the faunal divide. Figure 23 shows the region (together with a variant of Wallace's line) as clearly lying midway between the continental bases of Asia and Australia.

A faunal interchange in both border districts from the early

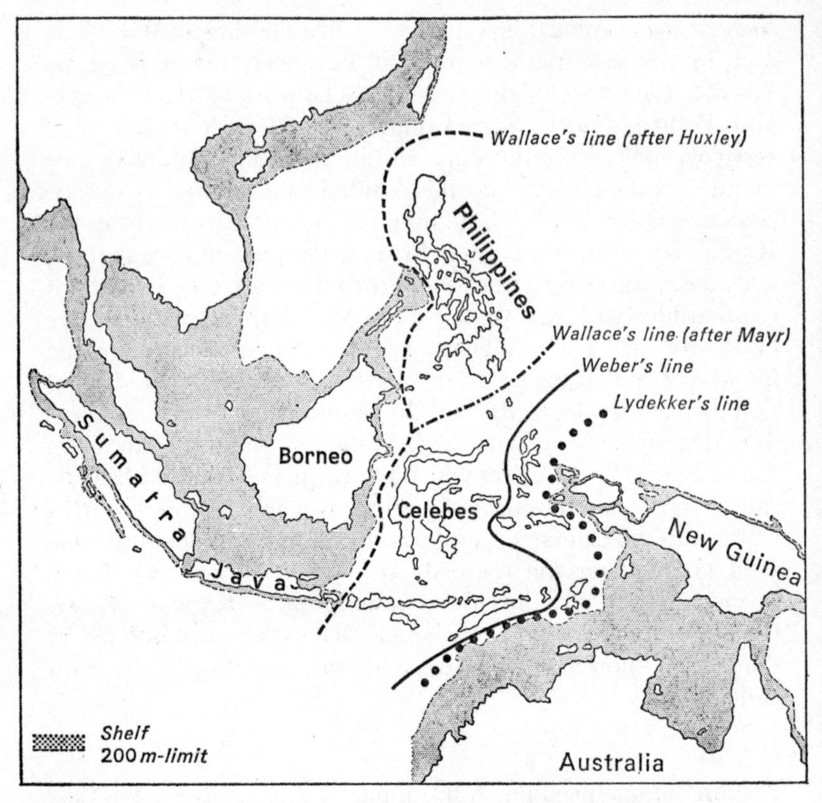

Figure 23 Faunistic boundaries in Wallacea (after MAYR, 1944)

Tertiary onwards is to be expected. According to the theory of continental displacement, Australia first drifted to this position in the late Mesozoic, whereas the area connecting the region with the Indo-Malayan zone is essentially older. Within Wallacea itself, faunal interchange across Pleistocene land-bridges took place between the Celebes and the Sunda Islands, and between the latter and the Lesser Sunda Islands (perhaps even with northern Australia). There were also transitory land connections during this epoch with Borneo and Australia. Lydekker's line remains problematical; strong relationships exist between Wallacea and the Papuan region (see p. 80).

Genera endemic to this region are the hooved deer-pigs or

babiruses (*Babirussa*) and the black ape (*Cynopithecus*). Numerous species from Oriental groups are also endemic, and the avifauna in particular is richly divided into species and superspecies; these provide a faunal identity for single islands. The bird fauna consequently indicates particularly clearly the finer points of the faunistic structure of the region, although other groups are indicative too. The stepwise penetration of Oriental and Australian faunal elements is particularly well seen in the chain of the Lesser Sunda Islands. The faunal jump between Lombok and Bali is especially great even though the two islands lie only some 30 km from one another (see principle 2, p. 49). Wallace's line, therefore, passes between these two islands; here, WALLACE discovered, a journey from Bali to Lombok was like travelling biologically from Asia to Australia, an impression later confirmed by many other zoologists, HAECKEL for example.

Bird and reptile species of the Lesser Sunda Islands
(Oriental: Australian components as per cent)

	Bali	*Lombok*	*Sumbawa*	*Flores*	*Alor*
Birds	87:13	72:28	68:32	63:37	57:43
Reptiles	94:6	85:15	87:13	78:22	75:25

Neotropis

The subcontinent of South America together with Central America, the West Indies and the southern part of Mexico (southern Sonora) is designated the zoogeographical region of Neotropis or Neogaea. Of all the faunal regions of the world, this one is the most distinguished by its high level of individuality, for the flora and fauna contain many characteristic and endemic groups and relationships with other regions are relatively slight. South America has led an isolated insular existence for long geological periods and, as a result, this has enabled independent associations, often at the systematic rank of family, to form or be retained. Transitory terrestrial connections with the Nearctic were not renewed until the late Tertiary and Quaternary, when the immigration of elements of the Holarctic fauna was rendered possible, for example of the large carnivorous cats, the puma and jaguar.

The old elements of the Neotropical fauna are represented in particular by the primitive mammalian groups of edentates (Edentata) and marsupials (Marsupialia). At the latest, these must have immigrated from the Nearctic in the early Tertiary and before the land connection with the north broke. They then underwent further development in Neotropis (or became extinct there as did the Notungulata, primitive hooved animals). Other even older groups may have been of southern origin and, as Gondwanaland elements, been the descendants of groups previously of Antarctic distribution. A relict fauna of this nature which shows pronounced relationships with Notogaea and especially New Zealand (see p. 81) occurs particularly in south-western Neotropis in the montane areas of the southern Andes. On the other hand, there are several groups which show the close Mesozoic kinship of Neotropis and Africa, for example freshwater fish of the families Cichlidae and Characidae, boa constrictors (Boinae) (apart from Neotropis, these are found elsewhere only in Madagascar), and even perhaps the iguanid lizards (Iguanidae). Thus, whereas the fauna of the Andean mountains shows old relationships with New Zealand, the fauna of the tropical lowlands is more closely related to Africa. This fact is not in full accord with the suggested model of the original Gondwanaland continent (figure 16), although it does agree with a recent variation of this proposed by HARRINGTON (1965). According to this variation, only the eastern part of Neotropis belonged to the former Gondwanaland, and the Andes together with part of Antarctica and New Zealand represent the inwardly compressed remnants of a circum-Pacific land girdle (see p. 81 and figure 28).

The fauna of the Antilles and Central America displays considerable individuality, although Tertiary land-bridges between the islands and the mainland must have existed. The endemic family of solenodons (Solenodontidae), a very old insectivoran group, indicates the level of isolation attained in this area.

There are a number of distinctive animal groups endemic to Neotropis and they give the region a special and distinct zoo-

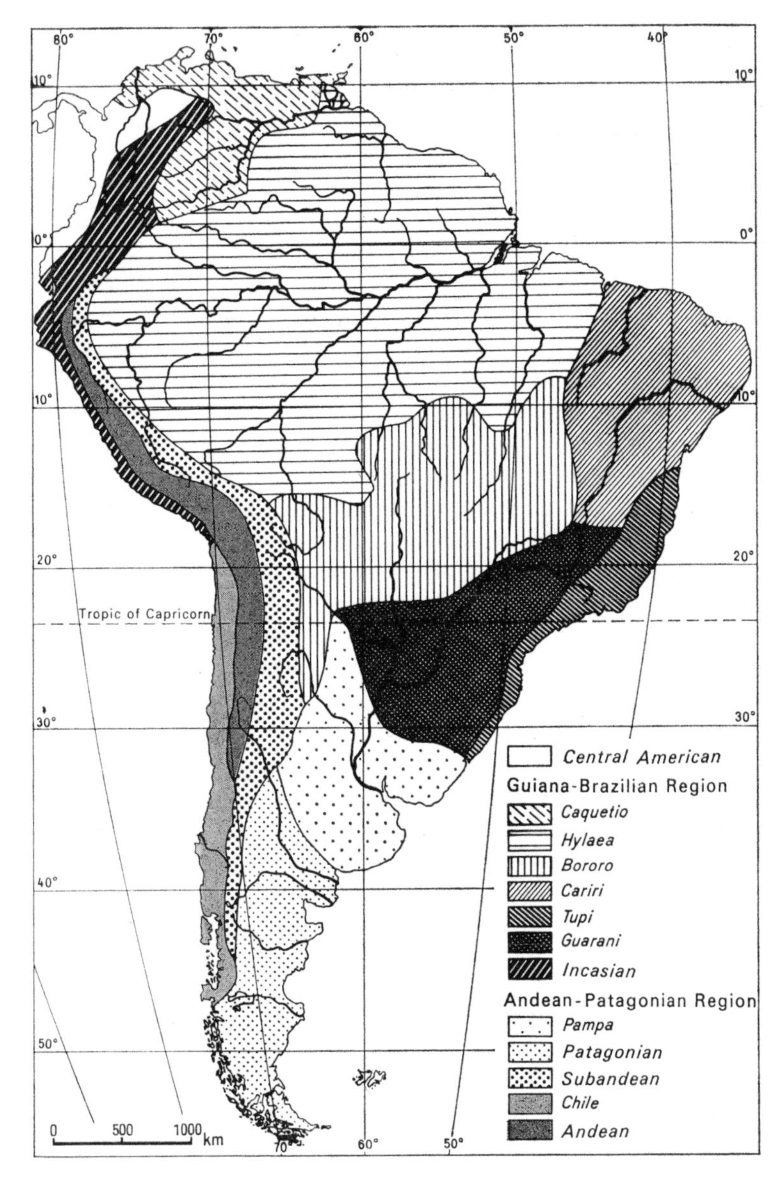

Figure 24 Neotropical zoogeographical regions (after FITTKAU, 1969)

geographical position. Among mammals there are the very old and primitive marsupials belonging to the opossum rat family (Caenolestidae) and marsupial rats (Didelphidae), as well as the edentates with three extant families, the sloths (Bradypodidae), the armidillos (Dasypodidae) and the anteaters (Myrmecophagidae). (It may be noted, however, that the armidillos extend as far as Mexico, and one species of marsupial rat extends well into North America). The rodents also provide distinctive families: certain New World porcupines (Erethizontidae), the guinea-pigs (Caviidae), viscachas and chinchillas (Chinchillidae) and agutis (Dasyproctidae). The artiodactyls are represented by the endemic camel–antelope group or llamas, and the bats by as many as six special families. Finally, with regard to primates, New World monkeys (Platyrrhina) have been present since Miocene times. Endemism is also extensive amongst the avifauna: of the 93 families which occur in Neotropis about a third are endemic to the region, including the ovenbirds (Furnariidae), rheas (Rheidae), screamers (Anhimidae), tinamous (Tinamidae), hoatzins (Opistocomidae) and cotingas (Cotingidae). Besides those Neotropical–Ethiopian families of freshwater fish previously mentioned, the mailed catfishes (Callichthyidae) and the Loricaridae should be noted as particularly distinctive forms.

The fauna is further supplemented by 'pan-American' groups – groups which presumably were endemic to Neotropis originally but which now also occur in the Nearctic. The peccaries (Tayassuidae) and racoons (Procyonidae) provide examples, as does also, strictly speaking, the opossum rat already referred to. Birds of this group include the humming birds (Trochilidae) and the tyrant flycatchers (Tyrannidae) in particular.

In the faunistic subdivision of the region, Central America and the West Indies clearly rate as a special subregion, and all biogeographers agree that at least an Andean–Patagonian and a Guianan–Brazilian fauna can be differentiated. Well-studied groups clearly indicate the nature of the zoogeograpical substructure of the region, as represented in figure 24 for mammals. Deviations in other groups are insignificant so that this zoogeographical pattern has a general validity. The Hylaea to

Guarani subregions include the tropical lowlands' zone and are characterised above all by the central rainforest and its faunal elements. Sharply delineated from these elements are the Andean elements of the western mountains (Incasian and Andean–sub-Andean) and adjacent savannahs (Pampa, Patagonian). These display many relationships to Notogaea (see below). The mountainous parts of the Caquetio and Incasian subregions form a local developmental centre and possess numerous endemics.

The Galapagos Islands also belong to the Neotropical zoogeographical region, although they were colonised by a singular and much reduced fauna. Their reptile fauna, in particular, contains many notable relicts with Mesozoic or early-Tertiary regional connections to Neotropis: the large, vegetarian iguanas such as the marine and terrestrial lizards (*Amblyrhynchus, Conolophus*), the geckos and the snakes and remaining lizards – each in an endemic genus (*Dromicus, Tropidurus*). The mammals of the islands (a rat and a bat) are either recent immigrants or have been introduced. Of the birds present, Darwin's finches (Geospizinae) have undergone adaptive radiation into numerous species, and as such provide a famous example of rapid evolution in an immigrant species (see p. 85).

Notogaea

Australia and the outlying continental islands of Tasmania and New Guinea, together with New Zealand and the Oceanic islands of the Pacific, comprise the zoogeographical region of Notogaea (figure 25). Its various subregions, especially the oceanic island subregion, are characterised by an eastwards increase in the number of gaps in the fauna present, so that finally it is almost impossible to associate the fauna with any large faunal region. The Australian continent is of central significance; its fauna radiates in all directions, and only New Zealand, New Guinea and New Caledonia possess strongly endemic, independent features in addition to Australian elements. Other parts, particularly the Hawaiian islands, have a fauna whose origin has more an ecological than geological basis.

The old Notogaean elements stem from former terrestrial connections of the main part of the region with other continents. The Mesozoic relicts show a strong relationship with Neotropis, but, remarkably, a much weaker one with Africa, even though the Gondwanaland connection would suggest otherwise. Those groups distributed in Australia *and* South America are referred to as amphinotic groups, or, following HENNIG, as 'A–S' groups.

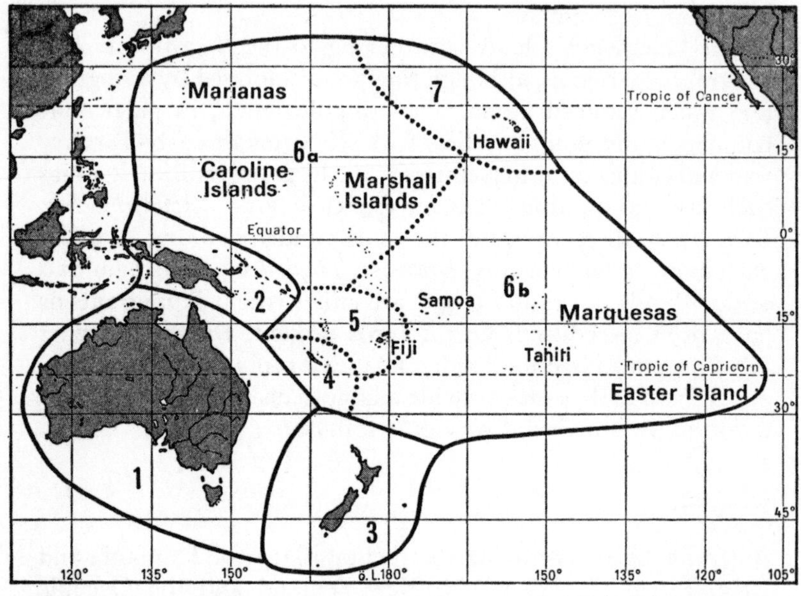

Figure 25 Division of Notogaea into zoogeographical regions (after GRESSITT, 1961). 1, Australia; 2, Papua; 3, New Zealand; 4, New Caledonia; 5, East Melanesia; 6a, Micronesia; 6b, East Polynesia; 7, Hawaii

They are either of northern hemisphere origin, and thus arrived in Notogaea across late Mesozoic connections with Asia (as did the monotremes and marsupials), or originated in the southern hemisphere and attained their modern distribution via Antarctica, as did, for example, many archaic insect families, including some bugs (Peloriidae), stoneflies (Gripopterygidae) and chironomids (Aphroteniinae).

Australia

The Australian faunal region comprises both the Australian mainland and the island of Tasmania. The previous remarks concerning the archaic elements of Notogaea are especially applicable to this region. After the Mesozoic, the area underwent a long period of isolation which persisted for almost the whole of the Tertiary and until the Pliocene. As a result of this isolation, the fauna exhibits a high grade of distinctiveness, and species and very often even genera are endemic in many groups of animals, for example insects, spiders, snails and particularly the vertebrates. This is often the case in the higher taxonomic categories too. Among mammals, there are only a few late immigrants, and these display close ties with Asia: rodents, bats and the dingo or wild dog.

The platypus (Ornithorhynchidae), belonging to the mammalian subclass Monotremata, is endemic, whereas the echidna (Echidnidae) in the same subclass also occurs in New Guinea. The subclass containing the marsupials (Marsupialia) has radiated into a rich spectrum of distinctive forms in many endemic families recalling the life-form types of higher mammals; marsupial cats (Dasyuridae), marsupial moles (Notoryctidae), marsupial badgers (Peramelidae), marsupial gliders (Phalangeridae), marsupial bears (Phascolomidae) and kangaroos (Macropodidae). (It should also be noted that several representatives have spread into the New Guinea subregion and even, in part, into Wallacea.) The avifauna is particularly rich in endemic groups, and it is not the least surprising that the parrots brought back from Australia in the seventeenth century gave the continent its first name, '*Terra psittacorum*'. Three endemic families of parrots occur: cockatoos (Cacatuidae), dwarf parrots (Cyclopsittacidae) and lorikeets (Loriidae). The following families of birds are also noteworthy: the primitive, ostrich-like emus (Dromiceiidae) and cassowaries (Casuariidae), the lyrebirds (Menuridae), mound-breeders (Megapodiidae), the honeyeaters (Meliphagidae) and flowerpeckers (Dicaeidae) which resemble the humming birds, scrub-birds (Atrichornithidae), bellmagpies (Cracticidae), mudnest-builders (Grallinidae) and bower-birds (Ptilonorhynchidae). (Many of these families

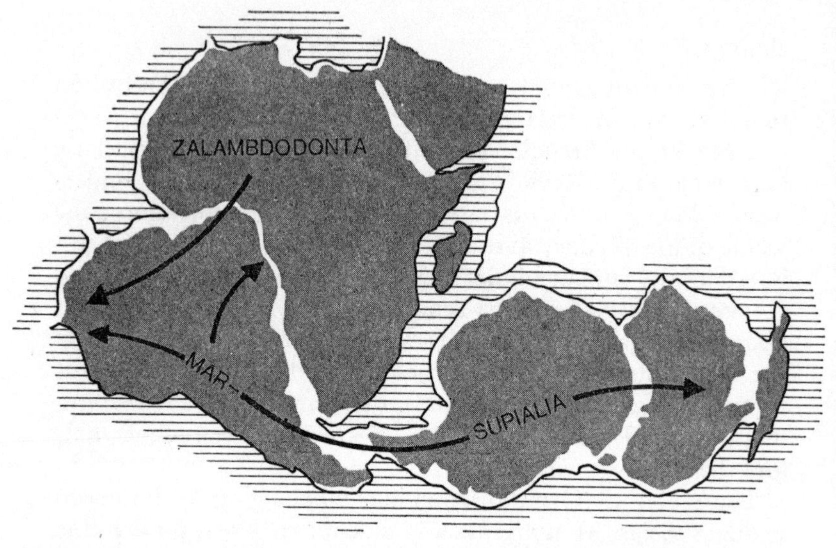

Figure 26 Scheme for the presumed areas of origin and migratory pathways of the marsupials and the alamiqui, golden mole and tenrec (Zalambdodonta) in the Mesozoic (after THENIUS, 1971)

also have single representatives in the Papuan region, and there are even single honeyeaters as far afield as Hawaii.) Of the reptiles, lizards of the family Pygopodidae and marine turtles in the family Carettochelyidae should be mentioned. Genuine freshwater fish are almost entirely absent, but native 'trout' (Galaxiidae) derived from marine immigrants of ancient Gondwanaland lineage are widely distributed.

Since European settlement, that is over a period of about 200 years, a secondary fauna has been introduced into Australia by man, and this, in part, has had a considerable influence on the faunal pattern and original elements. English rabbits (*Oryctolagus cuniculus*), which escaped from a farm in southern Victoria in 1859, have developed into the commonest mammal of the countryside and present a serious plague that can be controlled only by means of an artificially induced infection of the myxomatosis virus. The European fox (*Vulpes*), introduced for rabbit control, has also multiplied greatly and has had a devastating influence on the population density of the smaller endemic marsupials. In northern regions, the Asiatic buffalo (*Bubalus*), whose presence originated from earlier attempts at colonisation, has proliferated in a similar fashion.

Some basis for subdividing the continent into smaller zoogeographic areas is provided by the numerous east–west faunal disjunctions (figure 27), resulting from the extensive arid region in the middle of the continent which divides the country ecologically into two centres, a south-western one and an eastern one. The faunal and floral differences between these two centres

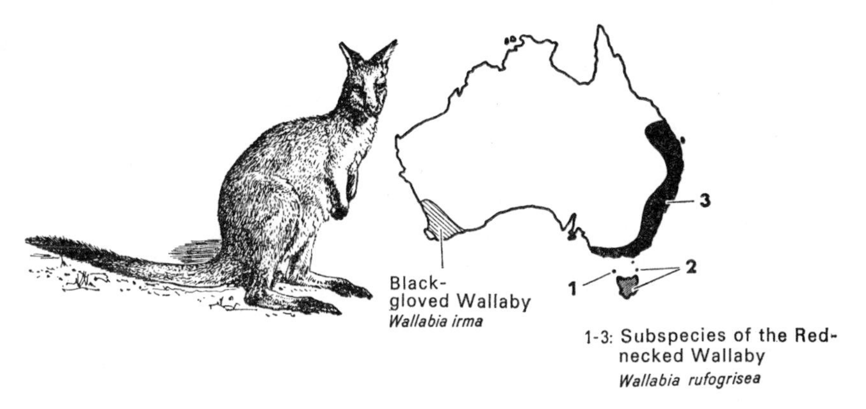

Figure 27 Disjunct distribution of two species of medium-sized kangaroos (wallabies) (after Marlow, 1965).

is so great that a complete isolation must be presumed to have taken place following a central Australian marine transgression. Among marsupials, for example, the honey possum (Tarsipedinae) is endemic to south-western Australia, and birds and invertebrates (insects, snails, spiders) also display considerable east–west faunal differences. Although Tasmania is now separated from the mainland by the Bass Strait, during the Pleistocene, transitory terrestrial connections existed as a result of lowered sea-levels; the last such connection occurred in the Würm. As a consequence, the Tasmanian fauna has close relationships to that of the mainland, and especially to that of the mountainous regions of Victoria. With regard to the mammals, most occur as endemic subspecies of east Australian species (see figure 27). Additionally, however, the isolation of the island has allowed some mammals to persist as relicts after they became extinct on the mainland in historical times. Consequently, a relatively rich marsupial fauna is encountered. The marsupial wolf (*Thylacinus*) and devil (*Sarcophilus*) are found only in Tasmania, for example (there is considerable doubt that the former genus is still extant). In the central region, several large lakes seem to have provided a special evolutionary centre for very old groups of the freshwater fauna, especially syncarid crustaceans of the order Anaspidacea (known since Carboniferous times).

New Guinea (Papua)

New Guinea and the immediately neighbouring islands of the Bismarck Archipelago and the Solomons comprise the Papuan subregion of Notogaea. The influence of elements derived from the Australian continent is marked here and determines the faunal picture. However, the precise evaluation of their origin remains difficult because the prevailing ecological relationships between the purely tropical islands and the subtropical and temperate parts of Australia are too dissimilar: the question nearly always remains open as to whether the Papuan elements represent relicts of a general Notogaean interglacial fauna that preferred warm conditions, or represent forms which originated endemically in this subregion. In addition to Australian elements, there are a number of Oriental species; these reached the Papuan zone in a strong wave of migration which took place during the Quaternary across the Sunda arc or via the Philippines and Mollucas. This wave renders precise delimitation from Wallacea problematical (Lydekker's line, figure 23). In contrast to the extensive Notogaean vertebrate fauna, the insects of the area show a close relationship to those of the Oriental region. This is so much the case that GRESSITT (1961) considered the Papuan area as an Oriental subregion. A verdict on this question remains outstanding.

Many of the animal groups which have been mentioned as endemic to Australia are also encountered as single groups of representatives in Papua: in the marsupials there are the gliding forms (Phalangerinae, cuscus and the sugar gliders) and the tree kangaroos (Dendrolaginae), and in the monotremes the long-beaked echidna (*Zaglossus*). There are several conspicuous bird groups which are so strongly concentrated in New Guinea that they can almost be considered as endemic there; only a few species of such groups occur in Australia itself, and then usually in northern Queensland. They include especially the birds of paradise (Paradisaeidae), the bellmagpies (Cracticinae) and the cassowary (Casuariidae).

New Zealand

There is no doubt that New Zealand and its subantarctic islands represent a distinct and special subregion of Notogaea. Although many groups of New Zealand animals display Australian affinities, there are no marsupials present, and, indeed, there are no endemic mammals here apart from two bat species. Moreover, old elements, especially such as occur in the mountains of the South Island, show a conspicuous relationship to the Neotropical fauna, and to the Andean–Patagonian fauna in particular. As BRUNDIN (1966) has proved for several insect groups (subfamilies of midges, Chironomidae), this Neotropical relationship may be even closer than any with Australia. It is possible therefore to differentiate between late Mesozoic, Gondwanaland relationships with Australia, South America and South Africa, and older connections which extended to the circum-Pacific geosyncline from Patagonia across west Antarctica to New Zealand. The latter only came into contact with the Gondwanaland elements later (figure 28). Which of these two types the old New Zealand elements belong to can only be ascertained after a careful analysis of systematic relationships, and is thus still unknown in most cases.

The presence of a relatively large number of endemic groups clearly sets the region apart. The avian fauna in particular – because predatory mammals are absent – has developed many unique ground-dwelling and flightless forms. They include the kiwis (Apterygidae), the moas (Aepyornithidae), and the owl-parrots (Strigopodidae). Especially noteworthy are the Aepyornithidae, for the family only became extinct within historical times. Some flying groups should also be mentioned: the kaka parrots (Nestorinae), New Zealand wrens (Xenicidae) and wat-

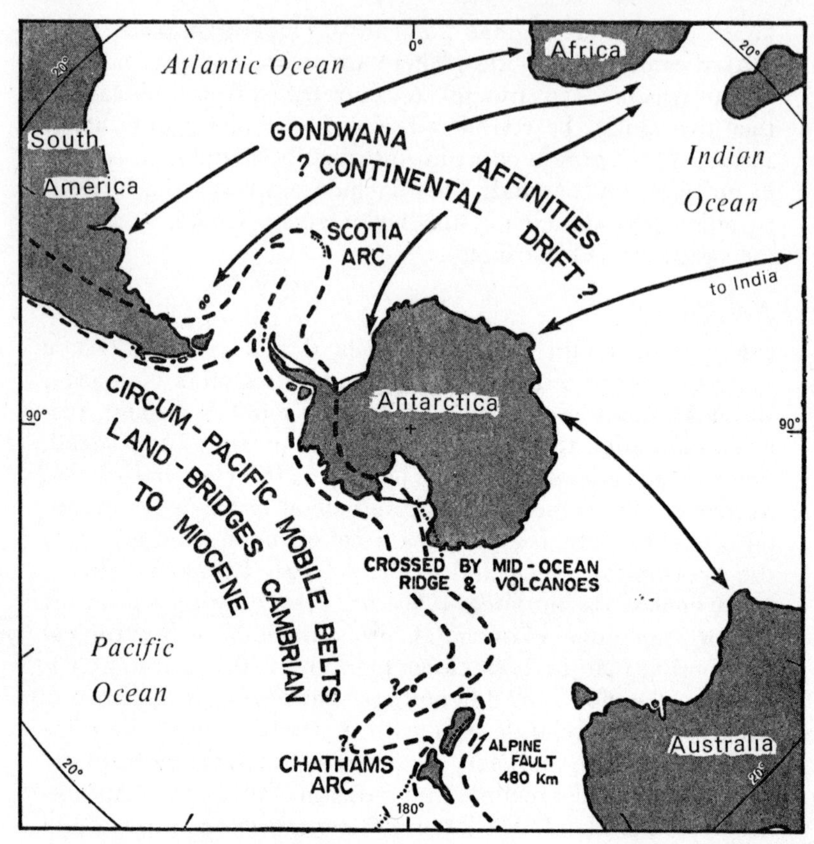

Figure 28 Palaeogeography of the southern hemisphere (after HARRINGTON, 1965). The time (Cretaceous to mid-Tertiary) shown is that when fragments of Gondwanaland drifted against the elastic and unresistant circum-Pacific belt thereby deforming it

tlebirds (Calaeidae). The outstanding reptilian example is provided by the tuatara or *Sphenodon* (Rhynchocephala), an early Mesozoic (Triassic) relict which has been able to survive because of its protected insular position; its sister groups, lizards, snakes and crocodiles, developed on continents. The frogs are represented by an archaic endemic family, the Leiopelmidae.

The secondary fauna that man introduced to New Zealand is now extremely widespread. Rabbits, goats and deer are so numerous in the forests that they have reached plague proportions and the hunting of them is officially encouraged. Wild pigs, chamois and ibexes, and various species of North American and Indian deer have also been introduced, as also have the Australian wallaby and 'possum' (*Pseudocheirus*), the latter now a forest pest. The European hedgehog and smaller carnivorous forms such as the fox and marten are other introduced species. Attempts to introduce zebras, llamas and guinea-pigs, to name but a few of many such species, were unsuccessful. The total secondary fauna numbers some fifty mammal species and numerous bird species, especially sparrows, finches, blackbirds and pheasants. This secondary fauna poses a serious threat to the native fauna of New Zealand, particularly its ground-dwelling birds, and certain sorts of parrot and the kiwi are already almost extinct.

Oceania

The fauna of the Pacific islands is characterised by a continuously increasing impoverishment eastwards. Old faunal elements are almost absent and isolated vertebrates (snakes, lizards and frogs) occur only in the geologically old continental regions (figure 25: 4 and 5). On those Polynesian islands of volcanic origin and consisting geologically of andesite, vertebrates are completely absent, except for birds, bats and flying foxes – all of course capable of flight. The insect fauna is descended from Oriental immigrants for the most part and is incomplete. On single islands the fauna has clearly resulted from colonisation by importation or immigration and is a function of the flight capability or dispersal activity of specific animal groups; it is directly related therefore to the distance of the island in question to the nearest mainland. The east Asiatic rat (*Rattus concolor*), which now lives on almost all Polynesian islands, was transported by ship.

New Caledonia (South Melanesia) and its neighbouring islands form an area characterised by a large number of endemic animals and plants. The island belongs to a continental bloc and possesses ancient biotic elements. In the Triassic, a transitory land connection with New Guinea and New Zealand presumably existed, the so-called inner Melanesian arc, and this finds expression in the fauna. Subsequently, long isolation led to the production of endemic genera in many groups, and among the birds there is even an endemic family, the kagu family (Rhynochetidae).

East Melanesia, that is those island groups from the New Hebrides to the Fijian islands, is distinctly more oceanic in its fauna than New Caledonia. Nevertheless, both geologically and faunistically, the Fijian islands do exhibit some continental features and their vertebrate fauna, for example, includes several species of frog in addition to snakes and lizards (the endemic iguanid genus *Brachylophus* should be noted). Furthermore, land snails belonging to *Placostylus* reach the eastern limit of their distribution here. In order to explain these continental faunal elements, some zoogeographers have advanced the idea that there was an outer Melanesian arc which formed a Mesozoic land-bridge with the mainland of Asia. The New Hebrides, on the other hand, are faunistically poorer and continental forms are almost completely absent; they were presumably first settled later. A relation to Samoa is evident, and the region is therefore extended by many zoogeographers to include this island.

Micronesia and east Polynesia are purely oceanic island regions, and in them faunal impoverishment is at its most extreme. Even terrestrial birds are scant here: of these there are in Samoa 33 species, in Tahiti 17, in the Marquesas 11 and on Pitcairn Island 4 species. The insects, likewise, are represented by diminishing numbers of families: the cicadas end in Samoa, and the long-horned beetles (Cerambycidae) in Tahiti; the weevils (Curculionidae), on the other hand, reach as far as Easter Island. However, apart from noting that the biota becomes more impoverished with increasing distance from the mainland, it is not possible to draw distinct borders within the region.

The Hawaiian Islands represent the extreme in faunistic isolation for their entire fauna is derived from Pleistocene or later immigrants. These colonised the islands by chance from forms transported by the wind or oceanic currents, and then developed into a distinctive fauna which, nevertheless, can be traced back to the ancestral elements of Oriental, Australian and American derivation. In this way, according to ZIMMERMAN's (1948) calculations, the 5000 or so insect species of the area (in about 100 families) have been derived from only 250 immigrant species and of these some have scarcely changed whereas others have radiated into numerous species and genera. This radiation can be followed particularly clearly in the fruit-fly genus *Drosophila*. H. L. K. CARSON (1971) suggests that the ancestral form immigrated to Hawaii some 700 000 years ago and then explosively radiated to the present stock of approximately 1000 endemic species. Considering the avifauna, one immigrant species from Neotropis gave rise to an endemic family, the Hawaiian honeycreepers (Drepanidae) with 9 genera and 22 species, and from Australian–Polynesian immigrant birds there has developed an endemic genus in each of the flycatcher (Muscicapidae) and honeyeater (Meliphagidae) families. Amongst terrestrial snails, two endemic Hawaiian families are derived from importations of this sort, the Achatinellidae and Amastridae.

Antarctica

The south polar region is usually disregarded in zoogeographical considerations of land faunas for it has scarcely any fauna worth noting because it is almost completely covered by ice. Recent detailed investigations, however, have stimulated interest in its zoogeography, and the results justify the position of Antarctica as a separate zoogeographical region.

Antarctica has undoubtedly been colonised during a large part of its geological history by a rich and partly tropical fauna and flora. Palaeozoic and especially Mesozoic (Triassic) rocks form the nucleus of the continent, and the discovery in 1968 of a fossil scaled anuran (Labyrinthodontia) from lower Triassic rocks provided the first proof of a tropical freshwater vertebrate fauna and confirmed the existence of an old Gondwanaland faunal connection (the Permo-Carboniferous *Glossopteris*-flora was also proved). In 1969 a further expedition found over 500 fossil bone fragments in the same formation. In these, besides amphibian fragments, reptile remains were dominant (Therapsida and Thecondonta), especially those of the predatory *Lystrosaurus murrayi*, the commonest fossil reptile of the lower Triassic African Karroo formation. Furthermore, according to COLBERT (1970), the small theriodonts (*Thrinaxodon liorhinus*) and cotylosaurs (*Procolophon trigoniceps*) that inhabited Antarctica were apparently the same species that existed in South Africa.

The extant terrestrial fauna comprises extremely cold-tolerant invertebrates, particularly mites and flightless insects. They occur on several nunataks (see p. 50) in the western region and especially on the coast of the Antarctic peninsula (Grahamland). The last enumeration (1965) of the fauna of the region listed 42 free-living species of mite, 12 species of insects and a freshwater crustacean (*Branchinecta*) of melt-water pools. If parasites of sea-birds and coastal marine mites are included, a total of about 150 arthropod species is arrived at. The most famous inhabitants of Antarctica are two species of penguin (the Emperor penguin and the Adelie penguin), but at least eight further species of breeding birds also occur (petrels, gulls, etc.). The penguins (Spheniscidae) are a very old group of birds, perhaps of Mesozoic vintage, which had their centre of origin in the Antarctic region and from there have radiated northwards in all directions.

The subdivision of the region reflects its geology. In the east, the Antarctic plate consists of an ancient nucleus of gneisses and granites over which a series of sediments has been deposited since the Silurian. It is these sediments which carry the Gondwanaland fossils. This flat part, constituting the greatest area of the continent, must have drifted to its present position after the Mesozoic disintegration of Gondwanaland (figures 26 and 28). As the connection with Australia was the last to be lost, the fauna probably resembled the Notogaean fauna until its final destruction by the Quaternary onset of cold conditions. The narrow western part of the continent, the Antarctic chain, consists of an alpine-folded chain of mountains, resembling in their construction the Patagonian Andes. This chain and its main anticline are part of a continuous geological formation which includes Grahamland, the Shetland Islands, South Orkney Islands, the South Sandwich Islands and South Georgia, and which is known as the Scotia Arc (figure 28).

According to one theory (HARRINGTON), the Antarctic chain was part of the circum-Pacific land belt which existed until the Miocene and of which Patagonia and New Zealand were also a part. BRUNDIN (1966) has advanced strong zoogeographical arguments for the theory (see p. 81). The fauna of western Antarctica at that time must therefore have resembled the Neotropical–Notogaean mountain fauna. The present polar position of the region was first attained when the land belt was forced inwards to its contemporary form by the drifting Gondwanaland Antarctic plate.

THE FAUNAL REGIONS OF THE SEA

In view of the fact that by far the greatest part of the earth's surface is covered by ocean, and that sea water represents the first animal habitat and was forsaken relatively late in the course of evolution, it is surprising that fewer than 10 per cent of all living animal species are marine. Of the more than one million extant species of animal, according to DE LATTIN, only some 85 000 multicellular species live in the sea. This relative paucity of marine species is all the more astonishing as a large number of animal groups are entirely or almost restricted to the sea. Such groups include amongst unicellular forms the radiolarians and the Foraminifera, and amongst multicellular forms especially Cnidaria (jelly-fish, sea anemones and corals), Echinodermata (sea-stars, sea-urchins, sea-cucumbers), many lower and higher 'worms' (Chaetognatha, Pogonophora, Nemertinea, Polychaeta), Cephalopoda (squids and octopuses), Tunicata (tunicates), Chondrichthyes (cartilaginous fishes: sharks and rays), the overwhelming number of orders and families of bony fish, and finally, among the mammals, seals (Pinnipedia), sea-cows (Sirenia) and whales (Cetacea). In addition, the majority of sponges (Porifera), snails and mussels (Mollusca) and crustaceans (Crustacea) are marine forms.

This disproportion in the marine fauna between the great variety of forms and the relatively slight amount of species diversity results from two distinctive environmental features, namely the uniformity of environmental factors and the lack of

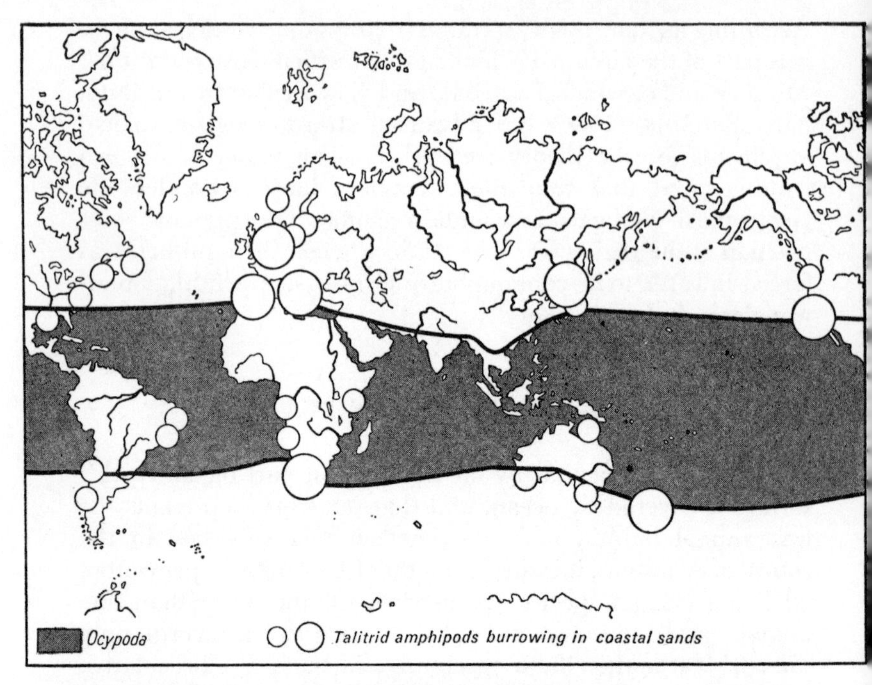

Figure 29 Distribution of the crab genus *Ocypoda* (tropical warm waters) and sand-burrowing amphipods (subtropical and temperate latitudes of both hemispheres) (after DAHL, 1952–53)

significant spatial division. There are scarcely any serious distributional barriers to forms which can swim or which live suspended in marine currents. Thus, speciation and radiation within small subregions that remained isolated for long periods has rarely been possible, and consequently the distribution of most species is wide, extends over whole climatic zones, or, as in many planktonic forms, for example, is even cosmopolitan. In contrast to this situation, the climatic diversity and greater opportunities for climatic and geographic isolation on land led to the development of much greater diversity in terrestrial forms, particularly amongst insects (which have more than 600 000 species).

Any subdivision of the marine fauna must be therefore more

a reflection of ecological than of geographical factors. Its gross subdivision, consequently, is into plankton, nekton and benthos communities (see p. 36) which, however, are arranged vertically in any place and so co-exist in a geographical sense. On account of the ecological significance of water temperature, a major spatial subdivision into an Arctic and Antarctic cold-water fauna and a tropical–subtropical warm-water fauna can also be distinguished. The cold-water fauna clearly has both Arctic and Antarctic elements (for example, the seals and marine birds), as well as some identical forms (bipolar disjunction, for example in many whales). Examples of exclusively tropical forms are given by coral reefs and their attendant fauna. The salinity differences of some marine areas is not of great zoogeographic significance for few marine forms can tolerate either an increase or a decrease in the salt concentration of their environment (figure 3).

The detailed faunal structure shows a series of small zoogeographical areas based on species and genera. Essentially, these areas follow the continental divisions. Whereas cold-water forms are almost always distributed in a circumpolar fashion, the warm-water fauna does show some distinct regional differences. The large continental blocks and ocean basins, in particular, form barriers running north–south and give rise to four subregions of the pelagial: the west Atlantic, the east Atlantic, the Indo-Pacific and the east Pacific. The east Pacific barrier has the most marked effect. Thus, those Indo-Pacific benthic species with pelagic larvae attain their eastern limits here, for after a certain developmental time the larvae must sink to the bottom and find a firm substrate. The littoral forms of the warm-water benthos are bound to the coast and are therefore isolated from one another by the continental blocs stretching southwards from cold-water regions. As a result, numerous vicarious pairs of species develop on both coasts of the American (American barrier) and African continent (African barrier).

With regard to the abyssal fauna, that is the fauna of the unlit depths of the sea below 200–600 m, these barriers lose their effectiveness; ecological conditions (water depth, temperature, nature of the bottom and feeding) come completely to the fore and it is scarcely possible to speak of abyssal zoogeographical divisions. Perhaps weak regional relationships occur as a result of the derivation of some deep-sea inhabitants from the littoral fauna. In any event, the extraordinary difficulty of deep-sea zoological investigations means that it is still not possible to recognise widely applicable faunistic principles. In the future, modern investigations using submarines and underwater television can be expected to provide essentially new information about the fauna of the deep sea.

ANIMAL MIGRATIONS

A number of animal species show regular migratory movements beyond the limits of their breeding areas so that their association with a specific biogeographical region is of only seasonal occurrence. Such periodic local changes are well known in many mammals, birds and fishes as well as in several insects. In part, they are natural phenomena that have long been recognised and recorded in legends and reports (locust plagues, bird movements, migration of rats, lemmings and so on). The reasons for these migrations, which often involve aggregations or large numbers of specimens, are of various sorts: reproduction or breeding, avoidance of winter cold, lack of food or overpopulation. Mostly, the physiological cause has been shown to result from endocrinological changes, and these in turn may be traced back to environmental influences or endogenous metabolic rhythms.

Insect movements are especially noticeable when they take the form of pestiferous plagues. Locust swarms in particular have been known as conspicuous natural phenomena since historical times (for example, as the Egyptian plague of the Bible). In North Africa recently, zoological observers estimated that in one locust swarm there was a total mass of 44 million tons, a value equivalent to several billion individual insects. Dragonflies, butterflies and beetles can also occur in huge swarms, presumably as a result of overpopulation, and these, being aerial, often leave the land and fly out over the open sea. Fatigued and dead individuals from such swarms may end as washed up coastal drift material that is sometimes kilometres long. One North American butterfly, the Monarch (*Danaus plexippus*), is a conspicuous seasonal migrant and oscillates annually in huge numbers between latitudes 32° and 48° N, so that the winter is spent in the south. Single individuals cover some 3000 km during the migration. In Europe, the large cabbage white (*Pieris brassicae*) is a migratory butterfly which also occurs in great swarms from time to time. In some years the moth *Laphygma exigua* migrates from the Caspian region to western Europe; during 1964, in a swarm numbering millions, such migrants were carried by air currents 3500 km in about fourteen days.

With regard to vertebrates, there are many migratory fish species in particular. The best known is the herring which, in swarms numbering thousands of millions of sexually ripe animals, migrates from the ocean into shallow coastal waters to spawn. Other migratory fish of economic significance include tuna, mackerel and cod; they, too, form migrating shoals which are followed by fishing fleets. Some species of fish live for most of the time in the sea but breed in rivers, and thus undergo migrations from the sea to estuaries and thence ascend rivers to suitable spawning beds (sturgeons, salmon). These *anadromous* migratory fish contrast with *catadromous* sorts which pass their growing period in fresh water and migrate seawards to spawn. The European eel (*Anguilla europaea*) develops from larvae which hatch in the Sargasso Sea (west Atlantic) and are then carried by the Gulf Stream to the European coast where they penetrate estuaries as elvers. Having ascended the rivers, the animals grow there until they reach the age of ten years, at which time, as sexually mature adults, they again migrate seawards (for an indication of the dependence of this migration on the moon's phases see figure 4). However, it remains an open question as to whether the European eel swims across the Atlantic and reaches the Sargasso Sea to lay eggs. According to TUCKER, they die prematurely, whereas adult American eels are able to migrate across the shorter ocean route and reach the spawning grounds. On the basis of this theory, therefore, all European eels are derived from American parents.

Birds exhibit particularly marked migratory activities, and there are very many species which leave their breeding grounds during the course of the year and undertake directional flight. Worldwide bird-banding experiments and results from numerous bird observatories give a detailed picture of the migratory pathways and speed of migration of certain species. The migration of the arctic tern (*Sterna macrura*) is the most amazing, for during the course of a year this species migrates from northern North America to the Antarctic. The longest and most rapid flight that has been observed is by the Pacific golden plover

Figure 30 Well-defined and narrow migration routes of the white stork into its overwintering areas (after SCHÜTZ, 1952)

(*Charadrius dominicus*): in 35 hours it flies from Alaska to Hawaii, a distance of 3300 km. There are also numerous observations concerning the annual migration of the white stork (*Ciconia alba*) from Europe to its overwintering area in southern Africa. The birds fly south along two narrow and quite fixed and separate routes, one including Gibraltar for western forms, and the other including Syria for eastern forms (figure 30). The actual navigation by migrating birds is an astonishing phenomenon and even now no fully satisfactory explanation of the ability is available. However, it is known from observations of carrier pigeons that radio-waves (transmitting masts) can interfere with the homing ability of these birds, and it has been shown experimentally in a planetarium that a certain nocturnally migrating warbler has the innate ability to navigate by the star patterns of the night sky. Apart from the regular, seasonal movements of migratory birds, there are also movements of an occasional, irregular evasive or dispersal sort. Thus the Siberian nutcracker (*Nucifraga caryocatactes*) occasionally reaches central Europe but only in particularly cold winters. And since 1938 the collared turtle-dove (*Streptopelia decaocta*) has immigrated from south-eastern Europe to central Europe where it is now widespread.

Among mammals, several bats undertake periodic migrations between their summer and winter quarters, which can be up to several hundred kilometres apart. In the hooved animals many grazing forms migrate to distant watering places or salt-licks, or migrate according to climatic or weather conditions. The reindeer is an example, and in summer great herds of this species follow the receding snow-line into the mountains. The irregular migrations of the lemming (*Lemmus lemmus*) are particularly famous. In some years, this rodent of the northern tundra region multiples so greatly that a population explosion occurs. An urge to migrate is then produced by the stress situation of the overpopulation, and as a result millions of individuals disperse from their breeding areas, undergo heavy mortality, and take part in a non-directional migration that ends catastrophically at the coast by drowning.

ANIMALS AND MAN

The relationship between man and animals is the subject of both ecology and the history of civilisation, for in many ways the very existence of man, in basic biological and economic terms, has been determined by this relationship. For discussion within a zoogeographical context, however, only a narrow sector will be selected from this broad spectrum, namely that having a direct geographical relevance: animals as the basis of man's existence, and man as a limiting or supporting environmental factor in faunal distributions.

Hunting

Human fossil remains indicate as clearly as does the behaviour of modern man that since the origin of *Homo sapiens* the human diet has consisted of both plant material and animal protein. Consequently, man has always been closely dependent upon animal sources of food, and, culturally, until a few thousand years ago man was a collector, hunter or subsistence fisherman. As food animals were not usually available in unrestricted fashion, they thus represented one factor regulating human distribution (see the Law of the Minimum on p. 7), and prehistoric hunting bands had either to search for areas of abundant wildlife or follow migrating herds. The nomadic Lapps are still at this cultural position, following half-wild reindeer on their seasonal migrations and traversing in the course of this some hundreds of kilometres.

Although elsewhere in nature there is a rhythmically fluctuating balance between predator and prey that renders possible continuing populations of both species (see the lynx and hare in figure 6), such a balance between man and his prey is endangered because of human technical superiority. Man is therefore the only carnivore able to exterminate its prey, because in this event a shift to other types of food (possibly plants) ensures that his own population remains unchanged. Even in early historical times certain animal species were hunted to extinction by man, the relatively sudden disappearance of the mammoth during the Ice Age probably being a case in point. And in New Zealand the giant flightless moas (*Euryapteryx gravis*) were exterminated in early historical time by the first human immigrants. Initially, almost inexhaustible stocks protected marine fish from serious depletion or extinction, despite the intensive fishing of coastal dwellers, but even the species in these stocks are no longer secure. Thus, with refined hunting techniques using explosive weapons, with changed human motivations (for example, the popularity of hunting as a sport and as a means of obtaining fashionable objects), and with

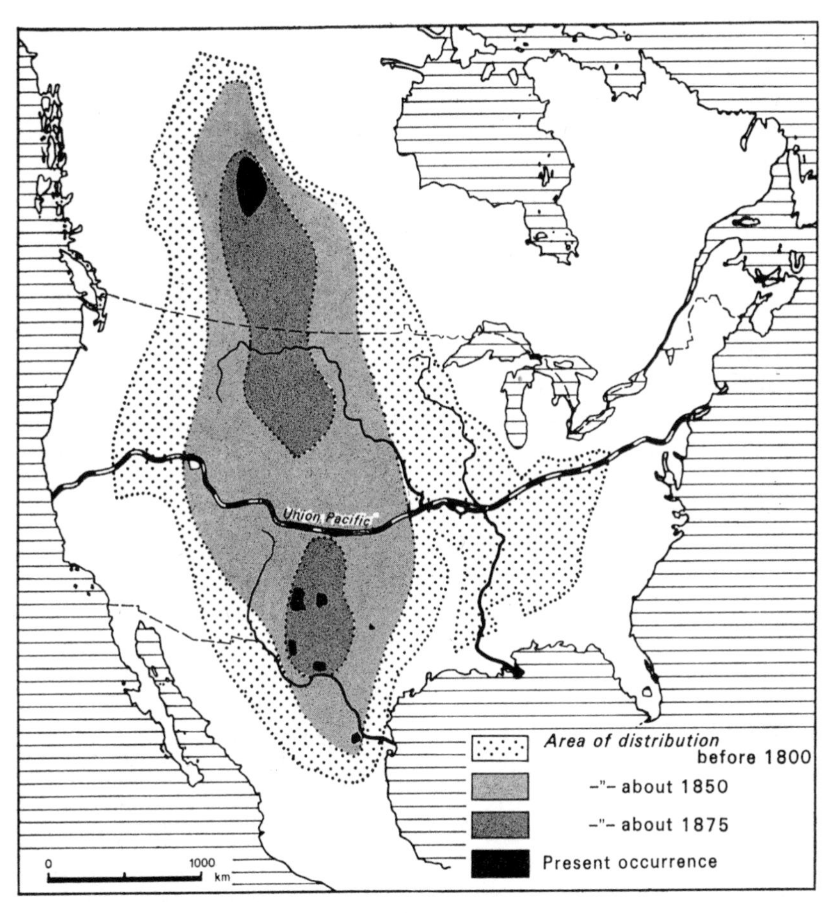

Figure 31 Former and present distribution of the bison in North America (after ZISWILER, 1965)

strongly increasing human population densities, human pressure on hunted animals has so intensified that the area of distribution of many animal species has become smaller very rapidly.

About 1700 the North American bison still had a population of some 60 million individuals in natural equilibrium with the relatively small Indian population. But following the construction of the Union-Pacific Railway and the hunting excursions this aided, the bison population changed so quickly during the nineteenth century that by 1850 only a few dozen examples of the bison still survived. The American passenger pigeon (*Ectopistes migratorius*), a species estimated to occur in single flocks of 2000 million individuals, was exterminated about the same time. Those animals on remote islands are especially threatened since in part they have no sort of flight instinct and have been used by seafarers from the Middle Ages onwards as a convenient source of provisions on voyages. In this way, the last example of the turkey-sized dodo (*Raphus cucullatus*) was slaughtered on Mauritius in 1681, and the last Steller's sea-cow (*Hydrodamalis gigas*) was killed on Bering Island in 1768. Even in Europe, several wild animals became extinct at a very early time as a result of intensive hunting; the European bison or auroch, for example, became extinct about 1627. Many other game animals have been exterminated from certain areas of their former distribution. Thus the lion became extinct in Greece in the early Middle Ages, in the Cape district about 1865 and in Morocco in 1922. The last bison in the Caucasus was killed in 1830, in Pennsylvania in 1825 and in Oregon in 1850. Other species, likewise strongly threatened, now exist only as a few individuals in zoological gardens or in conservation areas; the koala (*Phascolarctos cinereus*) provides an example, even though as recently as 1920 a million pelts of this species were subject to trade. According to information from conservation authorities, at the present time one additional mammal or bird species is exterminated per annum and hundreds are threatened.

On the other hand, it must be emphasized that in civilised countries populations of many species are now consciously maintained by official protective and conservation measures.

Some species which were already at the threshold of extinction have been preserved by strict protective regulations, by regeneration in zoos and by reintroductions, so that their numbers have again increased. (One study group of the United Nations, the World Wildlife Fund, has been entrusted with this task. International Red Books are published by means of which the organisation deals with the populations of all endangered species and maintains a watch on their development.) In addition to direct protective measures, many species of game animal have been established in and successfully colonised areas where they were not native. For instance, the trout (*Salmo trutta*) – a species particularly favoured as a sports fish – has been artificially introduced into rivers in almost all suitable mountainous regions of the world, and has become well established and developed extremely viable populations in many places (figure 32). The red deer (*Cervus elaphus*) has colonised the South American Andes and New Zealand so successfully that populations there produce greater and more powerful specimens than in their original area of distribution. (The extensive secondary fauna of Australia and New Zealand is discussed on pages 78 and 83.)

Domestic Animals

After the collecting and hunting phase and in early historical times, most of civilised mankind ceased being nomadic and settled in regions where plant and animal production in areas immediately surrounding settlements could provide sufficient sustenance for the populations therein. The cultivation of crops became necessary and was soon followed by the maintenance of domestic animals. The latter process was initiated by using enclosures, corrals and stables to prevent originally wild animals from free local movement, and to protect them from natural enemies and unfavourable climatic conditions. The elimination of natural selection and the introduction of artificial selection in the choice of reproductive mates then led to the appearance of domestication. The result was the formation and genetic fixation of particular domestic races which diverged from the wild race in certain desired characteristics and which

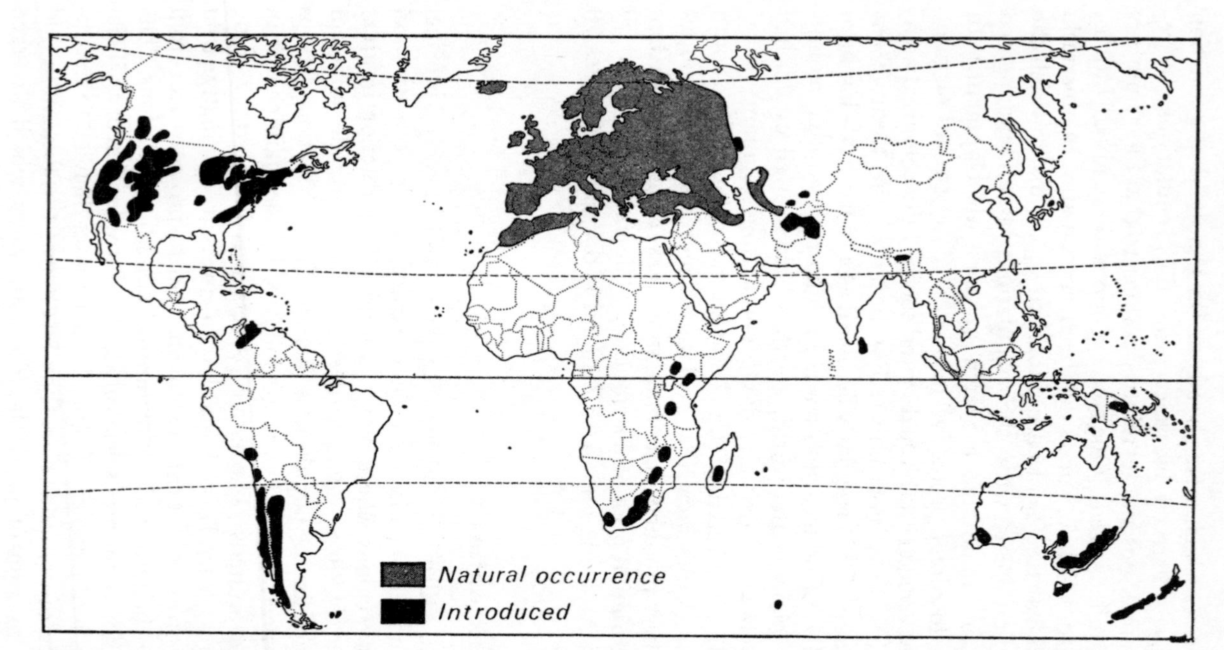

Figure 32 The original area of distribution of the brown trout and areas where it has been artificially naturalised (after MacCrimmon, 1970)

were very often larger and less deeply pigmented (bright pelts and white plumage).

Man's oldest domesticated animal, the dog, stems from the Middle Stone Age and as the domestic descendant of the wolf, its predatory behaviour led to its use in hunting. As draught animals the horse and ox substantially helped in agriculture, and their presence is demonstrable in prehistoric archaeological sites. Almost all animals domesticated later served as sources of milk or meat, although those providing pelts and hair were also important (see Table on p. 105) in the supply of material for clothing (leather, wool).

With the great increase in the world population of man during the past century and the recession in wild stocks of animals, pastoral activities have increased to such an extent that nowadays most existing large mammals are domesticated ones. Increasing rapidly like human populations, and about the same order of size, the total world sum of domestic animals has been: 1947/52, 2 352 600 000 animals; 1956/57, 2 741 600 000; and in 1962/63, 3 122 100 000 individuals (see Table on p. 106). These figures exclude cats, dogs, reindeer, yaks, llamas, alpacas, rabbits and poultry, and are based on a summary from the United Nations Food and Agriculture Organisation.

The process of domestication is still incomplete and breeding efforts continue constantly. Numerous domestic animals now exist in hundreds of races and breeds and are always being added to. Furthermore, efforts are being made to domesticate still more wild animals to serve as food or for agricultural purposes. The most recent attempt of this type relates to the musk ox (*Ovibus moschatus*). This animal is presently being domesticated in Labrador in special breeding areas set aside by the Canadian Government in order to develop a new resource for Eskimo populations of the region. The animals are extremely frugal, feed on tundra vegetation, are able to supply their own water requirements from snow and provide a very useful wool comparable with that obtained from Kashmir goats.

Fish culture must also be mentioned in this connection. The rearing of whitefish on special fish farms or in flooded rice fields has been known for centuries in east Asia, the first traditions of this technique dating from 2100 BC. And in Europe during the Middle Ages, the breeding of carp was of considerable importance, particularly in monasteries because of their need to provide special food during Lent. Today fish-farming is a branch of the economy in process of expansion. Even in classical Rome a marine fish, the Moray eel (*Muraena helena*), was kept in artificial ponds, and there is the story that for a triumphal dinner Julius Caesar procured 6000 of these fish from Hirtius the fish breeder, although it is a fable that they were fattened on slaves. Trout, especially the American rainbow trout (*Salmo irideus*) in addition to the native brook trout (*Salmo trutta*, figure 32), are being bred in increasing numbers on fish farms and smallholdings. Other fish also (pike, tench, pike-perch) are bred on fish farms in large numbers for release into artificial and natural waters. The Chinese grass carp (*Ctenopharyngodon idella*) represents the most recent enrichment of European fish culture; this species, unlike other pond fish, is able to feed on plants.

Man as a Supra-organic Factor

Man's effect on the distribution and abundance of animals has previously been largely confined to the results of hunting, direct killing, pastoralism and domestication. Now, however, modern technology has extended the effect of civilisation much further so that practically all life on earth is influenced. This process has taken place for the most part during the last century, but has accelerated catastrophically during the past decades of industrialisation. Human manipulation thus proves to be more significant than the limiting effects of natural factors (see p. 1), and the decisive factor for most species. THIENEMANN (1950) contrasted the influence of man as a 'supra-organic factor' with that of natural environmental conditions, and referred in particular to that branch of civil engineering concerned with dams, canals and so on, and which has had a marked and mostly deleterious effect on animals.

Those animals for which supra-organic factors represent a positive change in environmental conditions are referred to as synanthropes; in the man-made 'cliffs' of European cities, for example, several cliff-breeding species of birds have discovered new nesting possibilities as well as the necessary resources for a permanent association (swallow, swift, jackdaw, black redstart). Pigeons and sparrows also now form permanent and numerous populations in almost all large cities of the world, and they were introduced overseas by animal lovers for this purpose even, in part, during the nineteenth century. Human food supplies in granaries, barns and cellars are also the food supplies for many synanthropic rodents (rats, mice), and even inside houses, the wooden framework, furniture, carpets, food supplies and toilets represent the habitat of certain species, particularly insects with specialised food requirements. Such places also provide the breeding places of species exhibiting mass outbreaks (flies, moths, capricorn beetles, wheat beetles, dermestids, etc.). Gardens and parks attract and offer favourable environments for further species. In the case of the blackbird (*Turdus merula*), a shy forest-bird has become, in the course of just a few generations, a regular inhabitant of human settlements. The squirrel (*Sciurus vulgaris*), too, in many places now occurs more frequently in parks than in forests. Most especially, however, for all those animals ecologically associated with plants subject to agricultural or forestry monoculture, there now exists the potential for a population explosion and, in part, worldwide distribution.

At the same time, all these synanthropes have also become pests (vermin) to man. Thus rats and mice are poisoned in the same way as are moths, pests of grain, fruit and forest products, human parasites and disease vectors (midges, bugs, lice). Unfortunately, however, animals that are not pests but which live in the same habitat as pests may also be killed. Biological control aimed specifically at human pests and using the natural enemies of the pest involved attempts to avoid such undesirable side effects. How difficult it is to achieve the desired result by such interference with communities, and at the same time to maintain control over measures of this sort, is shown by the well-known example of the introduction of the European fox to combat the Australian rabbit plague (see p. 78).

On the other hand, the overwhelming majority of species avoid advancing human civilisation and are displaced by it.

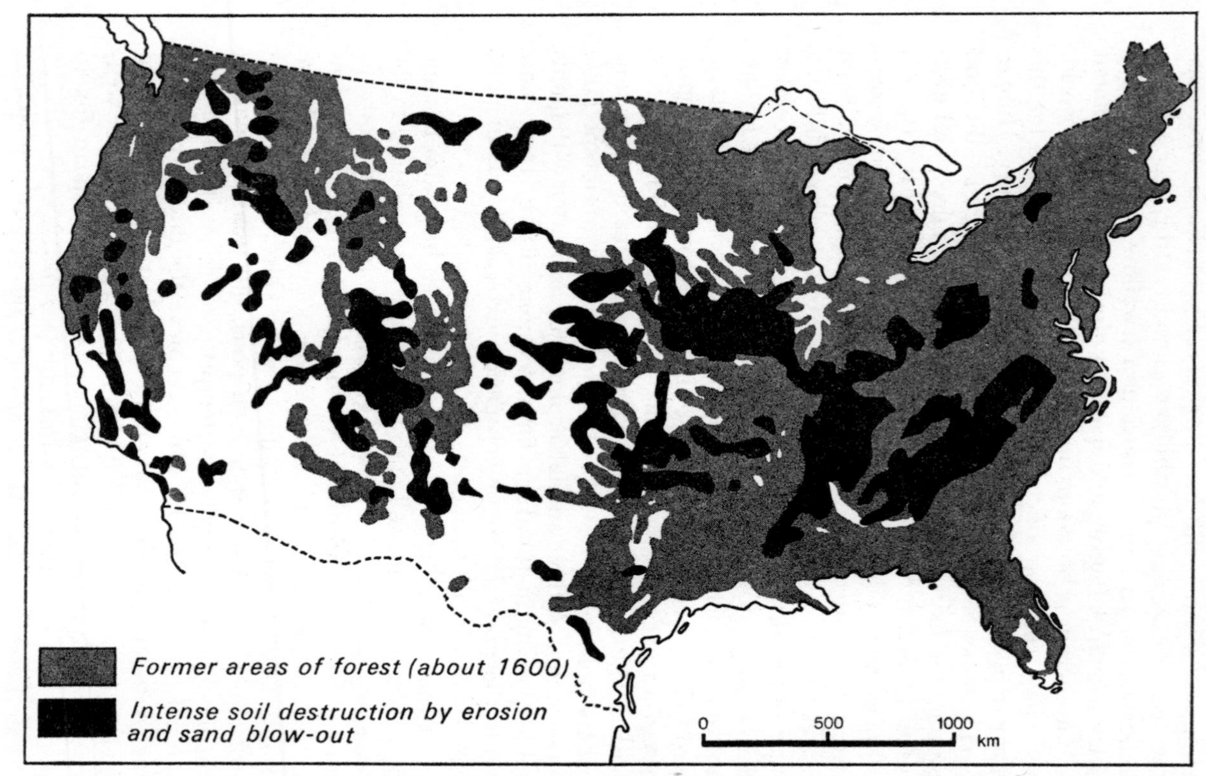

Figure 33 Forest destruction and change in faunal habitats through the effect of the human 'supra-organic factor' (after BERNHARD and GUTERSOHN, 1956)

Insofar as the displacement or extinction is of a direct nature (hunting, fishing, pest control) the subject has already been referred to, but mention should be made here of the indirect and increasingly important negative effects resulting from man-made environmental changes. The clearing of forests in particular has caused permanent remodelling of the world's landscape, a process that has taken place in Europe and Asia for thousands of years already, and in other places (especially the tropics) during the last century. The formation of grasslands (man-made steppes) follows deforestation, and with that the process of soil erosion commences. This, in time, may lead ultimately to the complete destruction of the vegetation layer. In this way, extensive areas now thickly populated by man have become uninhabitable for most of the fauna originally present.

Hydrological control structures markedly affect the marine and freshwater fauna in an overwhelmingly unfavourable manner. In any event, the ecology of running waters is significantly altered by canalisation, straightening and impoundment, and the presence of locks and dams will prevent the passage of migrating fish unless special fish passes are provided. Domestic and industrial waste water discharge resulting in water pollution causes further man-made devastation, and in most industrialised countries nowadays all medium-sized and large rivers, lakes and coastal waters are polluted. Hundreds of animal species (excluding fish, all the smaller benthic forms such as snails, mussels, crustaceans and insects) must now be regarded as having been displaced from these environments to a large extent. If they are not already extinct they are certainly in danger of becoming so.

Opposing these dangers, it should be noted that ideas concerning the protection of nature and environmental conservation have recently come more strongly to the fore. But it is only by energetic legislation and strict and drastic action that civilised countries can succeed in stopping the destruction of the landscape and the extinction of their fauna. For success, immense financial efforts are necessary, directed particularly towards the construction of water purification plants. Moreover, with regard to remaining primitive natural areas, a rethinking is needed which will lead from a ruthless exploitation of land towards a responsible and ecologically more sensible conduct. In setting aside nature conservation areas, national parks and similar regions, modern civilisation has the chance to demonstrate its power as a supra-organic factor in the service of the conservation of natural diversity and beauty of the biosphere.

The origin of the most important domestic animals (from NACHTSHEIM 1949)

Domestic animal	*Wild form*	*Distribution of wild form*	*Area of domestication*	*Time of domestication*
Dog	Wolf (and jackal?)	Northern hemisphere	Europe, Africa	Middle Stone Age (16 000–6000 BC)
Cow	Aurochs (extinct)	Old World	North Africa, India	Late Stone Age (6000–2000 BC)
Pig	1. European wild pig	1. Central and northern Europe	1. Baltic area	,,
	2. Mediterranean pig	2. Mediterranean countries	2. Mediterranean countries	,,
	3. Asiatic banded pig	3. South and East Asia	3. East Asia	,,
Sheep	1. Urial	1. Caspian Sea to the Himalayas	1. Southern Asia	,,
	2. Moufflon	2. Central and southern Europe, Asia Minor	2. Southern Europe	,,
	3. Argali	3. Central Asia	3. Central Asia	,,
Goat	1. Bezoar goat	1. Crete, Asia Minor, Caucasus, India	1. Southern Asia	,,
	2. Screw-horn goat	2. Caucasus, Asian highlands	2. Persia	,,
	3. Extinct goats	3. Eastern central Europe	3. Southern Europe	,,
Donkey	1. Nubian wild donkey	1. North Africa	1. Egypt, Ethiopia	,,
	2. Somalian wild donkey	2. Central Africa	2. Central Africa	,,
Horse	1. Przewalski's horse	1. Central Asia	1. Asia	,,
	2. Tarpan (extinct)	2. Southern Russia	2. Europe	,,
Cat	Yellow cat	Nubian area, Egypt	Spain, Italy, Egypt	*c.* 2000 BC
Rabbit	Wild rabbit	South-western Europe	France	Early antiquity
Pigeon	Rock-dove	Western Europe, Mediterranean area, Asia	Mediterranean area	*c.* 3000 BC
Hen	Red jungle fowl	India	India	*c.* 2000 BC
Goose	Wild goose	Europe, Asia	Southern Europe	Antiquity
Duck	Wild duck	Northern hemisphere	Europe	,,

Populations of the most important domesticated animals in terms of millions of individuals (from FELS, 1967)

	Cattle	Pigs	Sheep	Goats	Horses	Mules	Donkeys	Buffaloes	Camels
Europe									
1947/52	99.6	69.4	120.5	24.2	16.9	2.1	2.9	0.6	–
1956/57	108	98.8	129	17.7	14.3	2	2.8	0.6	–
1962/63	117.9	109.2	133.2	14.8	10.6	1.9	2.6	0.5	–
U.S.S.R.									
1947/52	55.8	19.7	76.9	15.6	12.8	–	0.8	–	0.3
1956/57	61	40.8	108	11.6	12.4	–	0·9	–	0·3
1962/63	86.9	69.9	139.6	6.7	9.4	–	0.8	–	0.3
Asia									
1947/52	245.2	95.1	156	128.9	11.9	2.6	17.9	80	3.1
1956/57	268	128.5	195	172.4	13.7	2.7	17.8	87.6	3.2
1962/63	301.7	218.9	223	176.8	13.8	2.8	19.4	99.3	3
Africa									
1947/52	91.6	4.1	121	87.2	3	1.6	8.7	1.6	6.6
1956/57	108	4.3	135	99	3.4	1.8	10.2	1.8	7.2
1962/63	120.9	5.3	138.8	109	3.5	1.9	11.2	2.2	7.7
Australia/Oceania									
1947/52	19.7	1.9	145.4	0.2	1.3	–	–	–	–
1956/57	23	2.2	192	0.2	0.9	–	–	–	–
1962/63	25.6	2.4	208.8	0.2	0.7	–	–	–	–
North and Central America									
1947/52	114.7	76	39	12.7	12	4.3	3.1	–	–
1956/57	138	71.8	39	15.4	9.9	3.9	3.6	–	–
1962/63	160.4	82.8	37.8	16.6	9.7	3.1	3.3	–	–
South America									
1947/52	136.7	35.3	121.7	18.7	18.1	4.3	3.3	–	–
1956/57	152	53.6	119	23.3	17.5	4.8	3.7	–	–
1962/63	166.5	65	122.3	26.9	16.6	5.6	4.3	–	–
Total									
1947/52	763.5	301.5	780.5	287.5	76	14.9	36.7	82.2	10
1956/57	858	400	917	339.6	72.1	15.2	39	90	10.7
1962/63	979.9	553.5	1003.5	351	64.3	15.3	41.6	102	11

Bibliography

BASIC WORKS AND TEXTBOOKS

COX, C. B., HEALEY, I. N. and MOORE, P. D., *Biogeography: an ecological and evolutionary approach*; Oxford 1973

DAHL, F., *Grundlagen einer ökologischen Tiergeographie*; Jena 1923

DARLINGTON, P. J., *Zoogeography. The geographical distribution of animals*; New York 1957

DE BEAUFORT, L. F., *Zoogeography of the land and inland waters*; London 1949

DE LATTIN, G., *Grundriss der Zoogeographie*; Jena 1967, Stuttgart 1968

GEORGE, W., *Animal geography*; London 1962

HESSE, R., *Tiergeographie*; Jena 1924

ILLIES, J., *Limnofauna Europaea*; Stuttgart 1967

JEANNEL, R., *La genèse des faunes terrestres*; Paris 1924

LEMÉE, G., *Précis de biogéographie*; Paris 1967

MARKUS, E., *Tiergeographie*; in: Handbuch der Geographischen Wissenschaft (Allgemeine Geographie 2); Potsdam 1933

NEILL, W. T., *The geography of life*; New York 1969

NEWBIGIN, M. T., *Plant and animal geography*; London 1968

REINIG, W. F., *Die Holarktis*; Jena 1937

RENSCH, B., *Die Verbreitung der Tiere im Raum*; in: Handbuch der Biologie vol. 5; Potsdam 1950

RUTTNER, F., *Grundriss der Limnologie*; Berlin 1952

SCHILDER, F. A., *Lehrbuch der allgemeinen Zoogeographie*; Jena 1956

SIMPSON, G., *The geography of evolution*; Philadelphia and New York 1965

THIENEMANN, A., *Verbreitungsgeschichte der Süsswassertierwelt Europas* (Die Binnengewässer 18); Stuttgart 1950

TISCHLER, W., *Synökologie der Landtiere*; Stuttgart 1955

SPECIAL REFERENCES

A good introduction to the zoogeography of the Holarctic, which also gives numerous details for the individual regions of the Nearctic and Palaearctic, is given by DE LATTIN, G., *Grundriss der Zoogeographie*; Jena 1967. For an account of the freshwater fauna of Europe see THIENEMANN, A., *Verbreitungsgeschichte der Süsswassertierwelt Europas*; Stuttgart 1950, and ILLIES, J., *Limnofauna Europaea*; Stuttgart 1967

A detailed modern account of the avifauna of Africa that also provides a general faunistic basis for the region is MOREAU, R. A., *The bird faunas of Africa and its islands*; London 1966

A modern, comprehensive account dealing with the Madagascan fauna is given by RICHARD-VINDARD, G., *Biogeography and ecology in Madagascar* (Monographiae Biologicae); Den Haag 1973

Information on the Sunda region is given by RENSCH, B., *Geschichte des Sundabogens*; Berlin 1936

A modern account for India is given in MANI, M. S., *Ecology and biogeography of India* (Monographiae Biologicae, in preparation)

Concerning Wallacea, a comprehensive study of lasting significance is RENSCH, B., *Geschichte des Sundabogens*; Berlin 1936. A modern supplement is MAYR, E., Wallace's line in the light of recent zoogeographic studies; in: *Q. Rev. Biol.,* 1944

Detailed information on the zoogeography of Neotropis is provided by the two volumes of FITTKAU, E. J., *et al. Biogeography and ecology in South America* (Monographiae Biologicae); Den Haag 1968 and 1969

Modern summaries for the whole of Notogaea scarcely exist, and older works are mostly to be rejected for they do not consider continental drift. A salient work, devoted to the central problem of dispersal, is that of BRUNDIN, L., *Transantarctic relationships and their significance, as evidenced by chironomid midges*; *K. svenska Vetensk. Akad. Handl.* (4th Ser.) **11** (1966), No. 1

Good information on Australia is given by KEAST, A., *et al. Biogeography and ecology in Australia* (Monographiae Biologicae); Den Haag 1959

Tasmania is discussed in WILLIAMS, W. D., *Biogeography and Ecology in Tasmania* (Monographiae Biologicae); Den Haag 1974

The voluminous special reports of the New York National History Museum: Archbold Expedition series, detail the results of zoological investigations of New Guinea. The insect fauna is dealt with by GRESSITT, J. L., in: Pacific Insects Monographs 2; Honolulu 1961

An excellent introduction to the palaeogeography and faunal history of New Zealand is given by FLEMING, C., *New Zealand biogeography, a palaeontologist's view*; in: Tuatara; Wellington 1962

An account of the zoogeography of Oceania and a survey of the analysis of single islands is given by GRESSITT, J. L., *Problems in the zoogeography of Pacific and Antarctic insects* (Pacific Insects Monograph 2); Honolulu 1961

The most recent detailed account of the biogeography and geology of the sixth continent (including also the work of H. I. HARRINGTON) is by VAN MIEGHEM, I.v., *et al. Biogeography and ecology in Antarctica* (Monographiae Biologicae); Den Haag 1965

The basic work on marine zoogeography is EKMAN, S., *Zoogeography of the sea*; London 1953

Animal migrations are comprehensively dealt with by HENDINGER, H., *Die Strassen der Tiere*; Braunschweig 1967

Endangered animals are listed and treated by ZISWILER, V., *Bedrohte und ausgerottete Tiere*; Berlin 1965

The most recent book on domestic animals and their origin is ZEUNER, F. E., *A history of domesticated animals*; London 1963

Index

Aardvark(s) 64, 66
Abiotic factors 1
Abyssal 37
Abyssal fauna 89
Abyssinian highlands 66
Achatinellidae 85
Acidity 13
Adelie penguin 86
Aepyornithidae 67, 81
Aerial plankton 9
Africa 13, 34, 52, 64, 67, 72, 76, 92
African barrier 89
African desert 35
African dormice 66
African jumping hares 66
African mole rats 66
African water hog 63, 67
Agamids 67
Agutis 74
Ailuridae 68
Alamiqui 78
Alaska 92
Alces 29, 53
Algae 8, 37
Alieni 17
Allan's rule 5
Alligator 60
Allochthonous ecosystems 24
Alopex 28
Alor 71
ALOSTA 45
Alpacas 99
Alpine hare 28
Alpine lakes 42
Alps 48, 50, 61
Amastridae 85
Amazonian crocodile 19
Amblyrhynchus 75
America 43, 45
American barrier 89
American eel 91
Amphibian(s) 31, 60, 86
Amphinotic groups 76
Amphipoda 39, 59, 88
Amphisbaenidae 64
Amphisbaen lizards 64
Anabiosis 3, 7
Anadromous fish 91
Anaspidacea 80
Ancient lakes 59
Ancylus 15
Andean fauna 75
Andean–Patagonian fauna 74, 81
Andean–sub-Andean fauna 75
Andes 72, 97
Anguilla 91
Anhimidae 74
Animals and man, relationships of 93
Annelids 37
Anomaluridae 66
Antarctica 72, 76, 81, 85, 86, 87, 92
Antarctic chain 86
Antarctic (marine) fauna 89
Antarctic tundral formation 28
Anteaters 74
Antelopes 33, 34, 63, 66
Antilles 72
Anuran 85
Apes 67
Aphids 19
Aphroteniinae 76
Apoda 31
Apollo butterfly 60
Apterygidae 81
Arachnocampa 9
Arctic fox 20
Arctic (marine) fauna 89
Arctic tern 92
Arctic tundral formation 28
Arctogaea 53, 54, 55
ARISTOTLE 43
Armadillos 74
Arrhenius equation 2
Artemia 12
Artiodactyls 74
Asellus 39
A–S groups 76
Asia 34, 57, 67, 69, 71, 76, 84, 99, 103
Asia Minor 50
Asitys 67
Assimilation 8
Atlantic 10
Atlas 48
Atolls 37
Atrichornithidae 77
Atriplectidae 39
AUGUSTINUS 45
Auklets 60
Aurelia 15

Aurochs 30, 34, 53, 96
Australia 5, 33, 37, 46, 69, 70, 71, 75, 77, 81, 86, 101
Autecology 1
Autochthonous ecosystems 24
Autumnal circulation 11
Avicennia 35
Azoic 11
Azure-winged magpie 58

Babiruses 70
Babirussa 70
Baboons 63
Bacteria 3, 41
Baikalidae 59
Balaenicipitidae 66
Bali 71
Balkans 48, 50, 61
Baltic amber 57
Baltic Sea 12
Bananas 32
Barbel 15
Barbus 15
Barnacles 37
Barrier reefs 12, 37
Bass Strait 79
Bathyergidae 66
Bathynellidae 59
Bathypelagial 36
Bats 8, 67, 74, 75, 77, 81, 83, 93
Baumberge 49
Bear(s) 30, 60, 66, 68
Beaver 30, 53, 60
Bed-bug 19
Beetles 13, 90, 101
Bellmagpies 77, 81
Benthal habitat 36
Benthic fauna 37, 39
Benthic organisms 13
Benthos 36, 41
Bergmann's rule 5
Beringia 55
Bering Island 96
Bering Straits 55
BERNHARD 102
BEURLEN 47
Bibos 63
Biocoenosis(-es) 22, 23, 24, 25, 26, 28
Biocoenotic principles 24
Biological control of pests 101
Biological systematics 48
Bioregions 36
of the continents 27
Biotic communitty 23
Biotic species 17
Biotope 16, 17, 23, 24, 25, 26
Birchir 66
Birds 4, 5, 8, 20, 25, 28, 30, 31, 32, 33, 34, 35, 60, 66, 67, 68, 71, 74, 77, 81, 83, 84, 86, 89, 90, 91, 96, 101
Birds of paradise 81
Birgus 6
Bischofsstadt 59
Bismarck Archipelago 80
Bison 34, 53, 95, 96
Bison 53
Bitterling 58
Black ape 70
Blackbird(s) 83, 101
Black redstart 101
Blue–green algae 3
Boa constrictors 72
Bogs 42
, inhabitants of 42
Boinae 72
Boreal 57
Boreo-alpine disjunction(s) 50, 57
Boreo-alpine forms 61
Boreo-alpine species 29
Borneo 69, 70
Boselaphus 63
Bovidae 60, 68
Bower-birds 77
Brachylophus 84
Brachypteraciidae 67
Brackish waters 12
Bradypodidae 74
Branchinecta 86
BRINCK 35
Brine shrimp 12
British Isles 46
Brooks 39
Brook trout 100
Brown bear 5
Brown trout 98
BRUNDIN 47, 81, 87
Bryozoans 37
Bubalus 79
Bucerotidae 64
Buffalo 63, 79
BUFFON 43
Bugs 21, 76
Bulbuls 64
Buprestidae 31
Bustards 33
Butterflies 8, 10, 19, 31, 90, 91
Bythinia 3

Cabbage white butterfly 91
Cacatuidae 77
Caddis flies 39
Caddis larvae 39
Caecilians 63, 64
Caenolestidae 74
Calaeidae 82
California 59
Callichthyidae 74
Camels 33, 55
Canada 28, 29
Canadian lynx 21
Canidae 60
Canis 53
Cape district 96
Capricorn beetles 101
Caquetio subregion 75
Carabus 60
Carboniferous 80
Carettochelyidae 78
Carnivores 18
Carp 39, 100
Carpathians 30, 61
Carrier pigeons 93
Carrion 19
eaters 18
CARSON 85
Caspian 91
Cassowaries 77, 81
Castor 53
Casuariidae 77, 81
Catadromous fish 91
Catarrhina 55

Catfish 66
Cats 99
Cattle 33, 60, 63
Caucasus 61, 96
Causal zoogeography 43, 48
Caves 6, 11, 24, 25, 38, 59
Cave salamander 59
Caviidae 74
Celebes 70
Central America 71, 72, 74
Central Europe 30
Cephalopoda 36, 87
Cerambycidae 84
Ceratotherium 63
Cervidae 60, 66, 68
Cervus 29, 53, 97
Cetacea 87
Chaetognatha 87
Chalcididae 31
Chameleons 64
Chamois 60, 68, 83
Chapin 65
Char 39
Characidae 72
Characteristic species 17
Charadrius 92
Cheetah(s) 33, 64
Chimpanzee 63
China 52, 68
China–eastern Himalayan district 68
China Sea 69
Chinchillas 74
Chinchillidae 74
Chinese grass carp 100
Chironomid(s) 3, 15, 39, 76, 81
Chironomus 15, 41
Chlorella 8
Chondrichthyes 87
Chorology 43
Chrysochloridae 66
Cicadas 9, 84
Cichlidae 64, 72
Ciconia 92
Ciliates 8, 38
Cimex 19
Citrus fruits 33
Civets 63, 67
Civettictis 63
Clausiliidae 50
Clothes moth 35
Cloud rats 68
Clunio 13
Cnidaria 87
Cockatoos 77
Cockchafer years 22
Coconut crab 6
Cocos Island 52
Cod 91
Coelenterata 6, 37
Coenographic 16
Colbert 86
Coliidae 66
Collared turtle-dove 93
Collembolans 9
Colugos 68
Colys 66
Comephoridae 59
Comoros 67
Competition 21, 22
Competitive exclusion principle 25
Condylura 55
Congo 66
Conies 66
Coniferous forest 29
Conolophus 75
Conservation 103
areas 104
Continental displacement 46, 47, 48, 70
Continental drift 46, 48, 55
Copepods 41
Coral reefs 3, 24, 25, 37, 89
Corals 37, 87
Corsica 61
Cotingas 74
Cotingidae 74
Cotylosaurs 86
Cracticidae 77
Cracticinae 81
Cranes 67
Crenon 38
Crested swifts 68
Cretaceous 82
Crocodiles 3, 60, 63, 82
Crustaceans 12, 37, 38, 59, 87
Cryptobranchus 60
Ctenopharyngodon 100
Cuckoo-rollers 67
Curculionidae 84
Current 9
Cuscus 69, 80
Cuvier 45
Cyanophyceans 41
Cyanopica 58
Cyclopsittacidae 77
Cynomys 55
Cynopithecus 70
Cypris 3

Dahl 88
Danaus 91
Darlington 46
Darwin 22, 45, 52
Darwin's finches 52, 75
Dasyhelea 3
Dasypodidae 74
Dasyproctidae 74
Dasyuridae 77
Date palms 32
Daubentoniidae 67
Death-watch beetles 7
Deciduous forests 24, 30
Deer 53, 60, 66, 68, 83
Deforestation 103
De Lattin 28, 44, 51, 56, 87
Dendrolaginae 81
Density 16
Dermestid beetles 101
Dermoptera 68
Desert animals 16
Deserts 34
Devil 80
Dew 34
Diatoms 41
Dicaeidae 77
Diceros 63
Didelphidae 74
Didermocerus 63
Dinaric west Balkans 61
Dingo 77
Discoglossid frogs 60
Discontinuity layer 41
Distributional history 43
Divers 60
Dodo 96
Dog 99

Domestic animals 97
, origins of 105
, populations of 105
Domestication 97, 99
Dominant species 17
Donkey 20
Draco 69
Dragonflies 90
Dragonfly larvae 39
Drepanidae 85
Dromiceiidae 77
Dromicus 75
Drosophila 85
Dwarf bushes 26
Dwarf parrots 77
Dystrophic lakes 42
Dytiscidae 39
Dytiscus 19

Eagles 33
Earthworms 6, 8
East Africa 33
East Asiatic rat 83
East Atlantic barrier 89
Easter Island 84
East Indies 68
East Melanesia 84
East Pacific barrier 89
East Polynesia 84
Echidna 77
Echidna 81
Echidnidae 77
Echinodermata 6, 37, 87
Echinosoricinae 68
Ecological animal geography 1
Ecological niche 23
Ecological valency 8, 14
Ecology 1, 24
Ecosystem(s) 23, 24, 26
Ectoparasites 21
Ectopistes 96
Edentata 72, 74
Eel larvae 10
Eel(s) 6, 13, 14
Eidmann 7
Elephant-bird 67
Elephant(s) 20, 63
Elephant shrews 66
Elephas 63
Elk 29, 53
Elvers 91
Emperor penguin 86
Emus 33, 77
Encysting 7
Endemics 53, 59
Endoparasites 21
Enemies (as factor) 20
English Channel 46
Environmental changes induced by man 103
Eocene 68
Epeirophoresis 46
Ephydra 37
Epilimnion 10, 41
Equidae 60
Equus 53
Erethizontidae 74
Erinaceidae 60
Eskimos 99
Estuaries 60
Ethiopian–Oriental regions 63
Ethiopian region 52, 57, 60, 63, 64
Ethiopis 65
Eunice 13
Eupelagial 36
Eurasia 28
Europe 4, 9, 30, 34, 50, 52, 57, 59, 61, 62, 63, 91, 92, 93, 96, 100, 103
European–East Asiatic disjunctions 58
European eel 91
Euryapteryx 94
Euryhaline organisms 12, 37
Euryoecious 15, 16
Euryphages 19
Eurythermous species 39
Eurytherms 4
Eustatic 38
Eutrophic lakes 41
Evolution 22
Exclusive species 17

Faroe Islands 20
Faunal regions 44, 53
Faunistics 43
Fels 10
Ferns 19
Fiddler crab 35
Field mice 20
Fijian Islands 84
Finches 60, 67, 83
Finfoots 63
Fish 4, 6, 9, 10, 18, 19, 35, 36, 37, 38, 39, 41, 59, 60, 66, 69, 72, 74, 78, 87, 90, 91, 94
culture 99
deaths 11
farming 100
Fittkau 73
Flat bogs 42
Flatworms 4, 19, 59
Fleas 21
Flores 71
Flowerpeckers 77
Flycatchers 85
Flying foxes 83
Flying lemurs 68
Flying lizard 69
Food (as factor) 17
Foraminifera 87
Forbes 46
Forel 40
Forest destruction 102
Formosa 68
Fox 5, 78, 83
Franz 24
Fresh water 11, 13
Freshwater fauna 103
Freshwater inhabitants 13
Freshwater limpet 15, 39
Freshwater organisms 38
Frogs 6, 9, 10, 82, 83, 84
Fruit-fly 85
Fumaroles 11
Furnariidae 74

Galaginae 66
Galago 63
Galapagos Islands 45, 52, 75
Galaxiidae 78
Game mammals 19, 34, 97
Gause's principle 25
Geckos 63, 75
Generic coefficient 25

Genesis 43, 45
Geospizinae 52, 75
Germany 49, 52
Gibbon 63, 68
Gibraltar 92
Giraffes 33, 52, 66
Glacial flea 15
Glacial mixed fauna 52
Glacial period 29
Gloger's rule 5
Glossopteris 85
Glowworms 9
Gnus 33, 66
Goats 83
Golden moles 66, 78
Gondwanaland 46, 47, 64, 68, 72, 76, 78, 81, 82, 85, 86
Antarctic plate 87
Gorilla 63
Grahamland 86
Grain beetles 7
Grain eaters 18
Grallinidae 77
Graphiuridae 66
Grassland 34
Grayling 39
Great apes 63
Greece 96
Greenland 28
GRESSITT 9, 80
Gripopterygidae 76
Grotta del Cane 11
Ground beetle 60
Groundwater(s) 38, 57
ecosystems 38
Guanas 67
Guarani subregion 74
Guianan–Brazilian fauna 74
Guinea-fowls 66
Guinea-pigs 33, 74, 83
Gulf of Bothnia 12
Gulf Stream 10, 91
Gulls 86
Gulo 53
GUTERSOHN 102
Gymnophiona 63, 64

Habitat 16
HAECKEL 1, 71
Hammerheads 66
Hard grasses 33
Hard waters 13
Hare(s) 5, 16, 20, 29, 53, 60, 67
Harpacticidae 59
HARRINGTON 72, 82, 87
Hawaii 78, 85, 92
Hawaiian honeycreepers 85
Hawaiian Islands 75
Heart-weight rule 5
Heaths 32
Hedgehogs 8, 60, 83
Heligoland 13
Heliornithidae 63
Hellenic–Balkans 61
Hemiprocnidae 68
HEMPEL 36
HENNIG 76
Herbivores 18
Herring 64, 91
Hesse's rule 5
Hibernation 5
Himalayas 48
Hippopotamidae 66
Hippopotamuses 66
Hippotraginae 66
HIRTIUS 100
Historical animal geography 43
Hoatzins 74
Holarctic(a) 55, 56, 57, 59, 60, 71
Holy apes of India 63
Homing ability of birds 93
Homo 94
Honeyeaters 77, 78, 85
Honey possum 79
HOOKER 46
Hooved deer-pigs 70
Hopping mice 33
Hormone cycle 14
Hornbills 64
Horses 20, 33, 53, 60, 67, 99
Hospites 17
Host–parasite balance 21
Hosts (of parasites) 21
Human populations 30, 96, 99
Humic material 13
Humidity 16
Humming birds 74
Humus 29, 30, 31, 42
Hungary 59
Hunting 94
HURLEY 47
HUSMANN 38
HUXLEY 45, 46, 53
Hyalodiscus 3
Hydrodamalis 96
Hydrogen sulphide 11
Hydrophilidae 39
Hydropsyche 10
Hydrozoans 8
Hyenas 33
Hylaea subregion, 74
Hylobatiinae 68
Hypertonic 38
Hypolimnion 41

Iberian peninsula 61
Ibexes 83
Ice Age 55, 59, 64, 68, 94
Iceland 50
Iguanas 75
Iguanidae 72
Iguanid lizards 72, 84
ILLIES 40, 61, 62
Incasian fauna 75
Incasian subregion 75
India 46, 52, 64, 67, 68
Indian bluebuck 63
Indian Ocean 67
Indifferent species 17
Indigenae 17
Indo-China 67, 68
Indo-Pacific barrier 89
Indridae 67
Ingolfiella 59
Inland waters 37
Inner Melanesian arc 84
Insect eaters 18
Insect larvae 19, 29, 41
Insects 9, 10, 19, 24, 28, 30, 31, 33, 37, 38, 42, 53, 76, 80, 83, 84, 85, 86, 88, 90, 101
International Red Books 97

Intestinal worms 21
Intra-specific competition 22
Irenidae 68
IRMSCHER 46
Irrigation 32
Islands 9
Isocoenoses 26, 39
Isotoma 15
Israel 59
Italy 61

Jackals 33
Jackdaw 101
Jaguar 71
Japan 57
Java 48
JEANNEL 46
Jelly-fish 15, 36, 64, 87
JENS 14
Jerboas 33
Jewel beetles 8, 31
Jordan's rule 25
Julius Caesar 100
Jungles 6, 24

Kagu 84
Kaka parrots 81
Kangaroos 33, 77
, disjunct distribution of 79
Kapuas River 69
Karroo formation 86
KASCHKAROW 27
Kashmir goats 99
Kerguelen Islands 10
Kidneys 13
Kiel Bay 24
Kilimanjaro 66
KIRCHNER 45
Kites 33
Kiwis 81
Koala 96
Korea 57
Krakatoa 48
KROGERUS 1
Kudu 63
KÜHNELT 28

Labrador 99
Labyrinthodontia 85
Ladybirds 19
Lagomorpha 60
Lake Baikal 59
Lake of Tiberias 59
Lake Ohrid 59, 61
Lake Prespa 59
Lakes 10, 24, 38, 40, 59
Lake Tahoe 59
Lake Tanganyika 64
Lake Toba 11, 41
Lake Victoria 13
LAMARCK 45
Laminifera 50
Land-bridges 45, 46
Laphygma 91
Lapps 94
La Rhune 50
Laurasia 47
Leafbirds 68
Leaf-monkeys 63
Leiopelmidae 82
Lemming(s) 28, 29, 90, 93
Lemmus 28, 93
Lemuria 67
Lemuridae 67
Leptosomatidae 67
Lepus 28, 53
Lesser Sunda Islands 70, 71
Lice 21
Lichens 26
LIEBIG–THIENEMANN's law of minimum 8
Life-form type 25
Light 8
Limnaea 4
Limnic organisms 38
Limnion 40
Limnophilidae 39
Limnoplankton 41
LINNAEUS 43
Lion(s) 33, 64, 96
Liponeura 15
Littoral areas 35
Liverfluke snail 4
Lizards 64, 67, 75, 78, 82, 83, 84
Llamas 55, 74, 83, 99
Locust plagues 90
Locust swarms 34
Lombok 71
Long-horned beetles 84
Lophiomyinae 66
Loricaridae 74
Loriidae 77
Lorikeets 77
Lorises 63, 66
Loxodonta 63
Lubomirksidae 59
Lumbricus 18
Lunar rhythm 14
Lungfish 66
Lydekker's line 69, 70, 80
Lynx 5, 20, 29, 30, 53
Lyrebirds 77
Lystrosaurus 86

MAACK 47
Macchia 32
MACCRIMMON 98
Mackerel 91
Macropodidae 77
Macroscelididae 66
Madagascar 67, 68, 72
rats 67
Mailed catfish 74
Makaham River 69
Malagasy 67
Malapteruridae 66
Malayan peninsula 69
Mammals 5, 8, 19, 20, 30, 31, 32, 46, 53, 55, 60, 66, 67, 72, 75, 77, 79, 83, 90, 93, 96, 99
, marine 28, 87
Mammoth 29, 94
Man 93, 94, 100, 103
, as a supra-organic factor 100
Mandrills 63
Maned rats 66
Mangrove swamps 35
Manioc 32
Manis 63
Margaritana 13
Marine currents 10
Marine fauna 87, 103
MARLOW 79
Marmota 55
Marquesas 84
Marshes 35
Marsupial badgers 77
Marsupial bears 77

Marsupial cats 77
Marsupial gliders 77
Marsupial moles 77
Marsupial rats 74
Marsupials (also Marsupialia) 69, 72, 76, 77, 78, 79, 80
Marsupial wolf 80
Marten(s) 53, 60, 83
Martes 53
Mascarenes 67
Mauritius 96
Mayfly 13, 25
larvae 39
MAYR 70
Mealworms 7
Mediterranean 32, 57
Megalobatrachus 60
Megapodiidae 77
Melanopus 59
Meliphagidae 77, 85
Melt-water pools 86
Menuridae 77
Merocoenoses 23, 24
Mesites 67
Mesoenanatidae 67
Mesozoic 47, 48, 55, 64, 67, 68, 70, 72, 75, 76, 77, 78, 81, 85, 86
land-bridge 84
relicts 76, 82
Metalimnion 41
Mexico 71, 74
Mice 19, 28, 67, 101
MICHAELSEN 46
Microclimate 2
Micronesia 84
Midge(s) 9, 13, 15, 19, 29, 81
Migration 90
Migration movements 13
Miocene 74, 87
Misgurnus 58
Mississippi 60
Mites 29, 86
Moas 81, 94
MOEBIUS 23
Moisture 6
Mole(s) 19, 55, 60
Mollucas 80
Molluscs (also Mollusca) 6, 38, 87
Monarch butterfly 91
Monard's principle 25
Monkeys 63, 67
Monophages 19
Monotremes 76, 77, 81
MOOLENGRAF 69
Moon rats 68
Moon's influence 13
Moray eel 100
Mormyridae 66
Morocco 96
Mosquitoes 9
Mosses 13, 19, 26
Moths 7, 9, 91, 101
Mound-breeders 77
Mudnest-builders 77
Mud-skipper 35
Mulberry tree 19
Munster 49
Muraena 100
Muscicapidae 85
Musk ox(en) 28, 29, 99
Musophagidae 66
Mussels 13, 37, 64, 87
Mustela 53
Mustelidae 60
Mutelidae 64
Myrmecophagidae 74
Myxomatosis 78
Myzopodidae 67
NACHTSHEIM 105
National parks 34, 104
Nearctic 53, 55, 57, 59, 60, 71, 72, 74
Nectariniidae 64
Necturus 59
Nekton 36, 41
Nematodes 29
Nemertinea 87
Neodendrocoelum 59
Neogaea 53, 71
Neotropical fauna 72, 81
Neotropis 64, 71, 72, 75, 76, 85
Nesomyidae 67
Nestorinae 81
Network of factors 16, 22
New Caledonia 75, 83, 84
New Guinea 75, 77, 80, 81, 84
New Hebrides 84
Newts 6
New World 43
apes 55
fauna 53
monkeys 74
porcupines 74
New Zealand 9, 72, 75, 81, 84, 87, 94, 97
wrens 81
Niche 23
Niphargus 38
North Africa 57, 90
North America 28, 30, 55, 57, 74, 92, 95
Notogaea 53, 72, 75, 77, 80
Notogaean fauna 86
Notoryctidae 77
Notungulata 72
Nucifraga 93
Numididae 66
Nunataks 50, 86
Oak–hornbeam forest 17
Ocean bottom 24
Oceania 83
Oceanic islands of Pacific 75
Octopuses 87
Ocypoda 88
ODUM 21
Ohridospongia 59
Okapi 52
Old World 43
apes 55
fauna 53
Oligocene 57
Oligochaetes 41
Oligophages 19
Oligotrophic lakes 41
Olives 32
Omnivores 18
Onychophora 31, 64
Opistocomidae 74
Opossum rats 72, 74
Orang utan 63
Oreal 28, 57
fauna 66
Oregon 96

Oriental fauna 68
Oriental region 57, 60, 63, 67
Ornithorhynchidae 77
Orycteropus 64
Oryctolagus 78
Osmotic balance 11
Ostracods 41
Ostriches 33, 66
Otter 30
Otter shrews 66
Outer Melanesian arc 84
Ovenbirds 74
Overturn (in lakes) 40
Ovibus 28, 99
Owl-parrots 81
Owls 8
Ox 99
Oxygen 10, 16
Oyster 23

Pachychilon 59
Pacific 9, 13, 37
golden plover 92
islands 83
Palaearctic 53, 55, 57, 59, 60, 63, 64, 68
Palaeogeographic statements 45
Palaeogeography 48
of southern hemisphere 82
Palaeozoic 47, 48, 64, 85
Palolo-worm 13
Pampa fauna 75
Pancarida 59
Pandas 68
Pangolins 63
Panotis 7
Panther 64
PARACELSUS 45
Paradisaeidae 81
Parasites 4, 19, 20, 21, 22
Parasitic hymenopterans 31
Parastenocaris 59
Parnassius 60
Parrots 63, 77
Papua 80
Papuan area 80
Papuan region 70, 78, 80
Passenger pigeon 34, 96
Patagonia 81, 87
Patagonian Andes 86
Patagonian fauna 75
Peat-bogs 24, 26
Peatmoss 42
Peccaries 74
Pedetidae 66
Pelagial 36
Peloriidae 76
Penguins 5, 86
Pennsylvania 96
Peramelidae 77
Perch 39
Periophthalmus 35
Peripatus 31, 64
Periphyton feeders 18
Permo-Carboniferous flora 85
Permo-Carboniferous period 46
Persian highlands 63
Pests 101
Petrels 86
Phalangeridae 77
Phalangerinae 80
Phascolarctos 96
Phascolomidae 77
Pheasants 83
Philepittidae 67
Philippines 68, 80
Phloeomyinae 68
Phoca 59
Phoeniculidae 66
Pholidota 63, 64
Phyllopods 41
Phylogenetics 48
Phytobenthos 37
Phytocoenosis 23, 26
Phytoplankton 41
Pieris 91
Pigeons 101
Pigmentation rule 5
Pigs 60, 63
Pike 20, 39, 60, 100
Pike-perch 100
Pine-heath forests 32
Pinnipedia 78
Pitcairn Island 84
Placostylus 84
Planaria 4, 5, 19, 49
Planarians 31
Plankton 10, 19, 36
Planktonic animals 13
Planktonic feeders 18
Plant juice feeders 18
Plant sociology 17
Platacanthomyinae 68
Platypus 77
Platyrrhina 55, 74
Pleistocene 57, 63, 69, 79, 84
glaciation 55
land-bridges 70
Pliocene 52, 77
Pogonophora 87
Polar bear(s) 5, 28, 29
Polar fox 28
Polar hare 28
Polluted waters 24
Pollution 103
Polycelis 4, 5
Polychaetes (and Polychaeta) 13, 37, 87
Polyhaline organisms 12
Polynesian islands 83
Polyodon 60
Polypedilum 3
Polyphages 19
Polypteroidei 66
Ponds 24, 29, 40
Pools 26, 40
Porifera 87
Possum 83
Potamochoerus 63, 67
Potamogalidae 66
Potamon 39
Poultry 99
Povilla 13
Prairie dogs 55
Predator–prey balance 20
Predators 20, 22
Preferent species 17
Presbytis 63
Pressure 16
Primates 74
Prionopidae 66
Procaviidae 66
Procolophon 86
Procyonidae 74
Proportion rule 5

Prop roots 35
Proteidae 59, 60
Proteus 59
Protopteridae 66
Protozoans (also Protozoa) 3, 6, 9, 41, 53
Psammal 39
Psephal 39
PSEUDOAUGUSTINUS 45
Pseudocheirus 83
Ptilonorhynchidae 77
Puma 71
Pycnonotidae 64
Pygopodidae 78
Pyrenees 50, 61
Pythons 31, 63

Quaternary 50, 57, 60, 71, 80, 86
Queensland 81

Rabbits 33, 78, 83, 99, 101
Racoons 74
Radiolarians 87
Rainbow trout 100
Rainforests 31
Rains of fish 9
Raised bogs 42
RAND 47
Ranges for some abiotic factors 15
Rangifer 28, 53
Raphus 96
Rats 33, 75, 90, 101
Rattus 83
Rays 87
Recessive species 17
Recolonisation of Krakatoa 48–49
Red deer 5, 29, 97
Reef-building corals 11
Refuges 50
Regional distribution of inland waters 40
Reindeer 28, 29, 53, 93, 94, 99
REMANE 12, 25, 26
RENSCH 46, 58, 69
Reptiles 30, 31, 67, 71, 75, 78, 86
Rheas 33, 74
Rheidae 74
Rheophilous fauna 39
Rheophilous species 10
Rhesus 63
Rhine 14
Rhinos (*Rhinoceros*) 63
Rhithron 39
  biocoenosis 40
Rhizophora 35
Rhodeus 58
Rhynchocephala 82
Rhynchomyinae 68
Rhynochetidae 84
Rice 32
Ring atolls 12
Ringed thrush 29, 50, 51
Riolus 13
Rivers 10, 13, 24, 38
Roan antelopes 66
Rodent(s) 16, 33, 34, 74, 77, 93, 101
Rotifers 3, 41, 53
Rove beetles 31
Ruffe–flounder region 17
Running waters 39
RÜTIMEYER 46
RUTTNER 42

Sable antelopes 66
Sagittariidae 66
Salamanders 60
Salines 37
Saline waters 37
Salinity 11, 16
Salmo 97, 100
Salmon 91
Salt lakes 24
Salt marshes 37
Salt springs 37
Samoa 84
Sand dunes 35
Sandpipers 35
Sarcophilus 80
Sargasso Sea 91
Savannahs 33
Scaly anteaters 63, 64
Scaly-tailed squirrels 66
Scarphirrhynchus 60
SCHÜTZ 92
Sciurus 101
SCLATER 45, 46, 67
Sclerophyll forests 32
Scopidae 66
Scotia Arc 86
Screamers 74
Scrub-birds 77
Sea 11, 12, 16, 24, 36
  anemones 87
  bottom 36
  , faunal regions of 87
  of Galilee 59
Sea-cows 87
Sea-cucumbers 87
Sea-feathers 37
Seals 5, 87, 89
  , freshwater 59
Sea-stars 87
Sea-urchins 87
Sea-weeds 37
Secondary fauna, of Australia 78
  , of New Zealand 83
Secretary birds 66
Serengeti 34
Seychelles 67
Sharks 87
Sheep 68
Shetlands 86
Shrew 19
Shrew-like rats 68
Siberia 5, 29, 55
Siberian nutcraker 93
Sibesia 49
Silk moth 19
Silurian 86
Simulium 10
Sirenia 87
Size rule 5
Slender loris of India 63
Sloths 74
Slugs 6, 31
Snails 3, 4, 37, 41, 59, 84, 85, 87
Snakes 67, 75, 82, 83, 84
Snoutfish 66
Snowshoe hare 21
Soft waters 13
Soil erosion 103
Solenodons 72
Solenodontidae 72
Solomons 80
Sonora 71
South Africa 81, 86

South America 46, 63, 64, 71, 76, 81
Southern continents 46
South Georgia 86
South Island (New Zealand) 81
South Melanesia 83
South Orkney Islands 86
South Sandwich Islands 86
Spain 5
Sparrows 83, 101
Sphagnum 42
Spheniscidae 86
Sphenodon 82
Spiders 9
Spiny dormice 68
Sponges 10, 37, 38, 59, 87
Spring circulation 11
Springs 10, 38, 39
Spring-tails 29
Squids 87
Squirrel 101
Staphylinidae 31
Steller's sea-cow 96
Stenoecious 15, 16
Stenohaline organisms 37
Stenophages 19
Stenophagous 19
Stenotherms 4
Steppes 33, 34
Sterna 92
Stoneflies 76
Stonefly larvae 39
Strangers (as species) 17
Stratification (of lakes) 13
Streams 10, 13, 39
Strepsiceros 63
Streptopelia 93
Strigopodidae 81
Struthionidae 66
Sturgeons 60, 91
Stygon 38
Sugar-gliders 80
Suidae 60
Sumatra 11, 41, 69
Sumatran River 69
Sumbawa 71
Summer stagnation (in lakes) 41
Sunbirds 64
Sunda 48
  arc 69, 80
  area 68
  islands 69, 70
Sundaland 69
Sus 63
Swallow 101
Swamps 24, 26, 35
Swarming movements 13
Swift 101
Synanthropes 4, 100, 101
Syncarida 59, 80
Synceros 63
Synecology 1
Syria 92
Syzygy 13

Tahiti 84
Taiga 28, 29
Talpa 55
Tarsiers 68
Tarsiidae 68
Tarsipedinae 79
Tasmania 75, 77, 79, 80
Tasmanian fauna 79
Tayassuidae 74
Temperature 2, 16
  tolerance limits 3
Tench 100
Tenebrionidae 7, 35
Tenrec 67, 78
Tenrecidae 67
Termites 9, 33
Terrestrial bioregions 26
Tertiary 50, 55, 63, 69, 71, 72, 75, 77, 82
  fauna 57, 60
  land-bridges 72
  refuges 59
  relicts 38, 59, 60
Teutoberg forest 49
Thalarctos 28
Thalasson 39
Thecodonta 86
THENIUS 78
Theory of permanence 45
Therapsida 86
Theriodonts 86
Thermal budgets 8
Thermal stratification 40, 41
Thermosbaena 57
THIENEMANN 1, 9, 24, 52, 69, 100
Three-toed woodpecker 56
Thrinaxodon 86
Thylacinus 80
Tibetan highlands 68
Tidal zones 13
Tiger 64
Tinamidae 74
Tinamous 74
TISCHLER 27
Toads 6
Touracos 66
Tree kangaroos 81
Tree shrews 68
Triassic 83, 85, 86
  relict 82
Trochilidae 74
Troglon 38
Trophic relationships (marine) 36
Tropical rainforests 31
Tropidurus 75
Trout 39, 97, 98, 100
Tuatara 82
Tubiculous polychaetes 37
Tubifex 41
Tubulidentata 66
TUCKER 91
Tuna 91
Tundra 24, 26, 28, 29, 52, 53
Tundral formation 28
Tunicata 87
Tupaiidae 68
Turdus 50, 51, 101
Turtles 31, 78
Tyrant flycatchers (Tyrannidae) 74
Tyrphobionts 42

Uca 35
UEXKÜLL 1
Ungulates 67

Urals 30, 63
Ursidae 60, 66, 68

Vacuoles 13
Vermin 101
Vicarious families 55
Vicarious genera 55, 63, 66
Vicarious species 53, 66, 89
Vicini 17
Victoria 78, 79
Viscachas 74
Viverra 63
Viverridae 67
VON IHERING 46
VOUK 14
Vulpes 78
Vultures 33

Waders 35
Wallabies 79, 83
WALLACE 45, 53, 69, 71
Wallacea 69, 77, 80
, faunistic boundaries in 70
Wallace's line 69, 71
Water beetle(s) 3, 19, 39
Water chemistry 13
Water lice 39
Water mites 3, 38, 39
Water pollution 103
Water rails 35
Wattlebirds 82
Waxwings 60
Weasel 53
Weather-fish 58
WEBER 69
Weber's line 69
Weevils 84
WEGENER 46, 47
West Atlantic barrier 89
West Indies 71, 74
Whale-billed storks 66
Whale(s) 20, 87, 89
Wheat beetles 101
Whitefish 99
White stork 92
Wild cat 5, 30
Wild dog 77
Wild horse 34
Wild pig 5, 19, 20, 83
Wind drift 9
Wind transport 9
Wolf(-ves) 5, 30, 33, 34, 53, 99
Wolverine 53
Wood-hoopoes 66
Woodpeckers 67
Wood-shrikes 66
Woolly rhinoceros 29
World Wildlife Fund 97
WUNDERLICH 47
Würm 79

Xenicidae 81

Yaks 99
Yangste Kiang 60

Zaglossus 81
Zalambolodonta 78
ZARATE 45
Zebras 33, 66, 83
Zebu 63
ZIMMERMAN 85
ZISWILER 95
Zonation of flatworms 4
Zoobenthos 37
Zoocoenoses 23, 26
Zoogeographical regions 45
of Ethiopia 65
of Europe 61, 62
of Neotropis 73
Zooplankton 41